AN INTRODUCTION TO THE METHODS OF OPTICAL CRYSTALLOGRAPHY

F. Donald Bloss

VIRGINIA POLYTECHNIC INSTITUTE AND STATE UNIVERSITY

Holt, Rinehart and Winston

NEW YORK   CHICAGO   SAN FRANCISCO   TORONTO   LONDON

FOR LOUISE

PREFACE

Some of the criticisms that have been directed at existing textbooks in optical crystallography have obvious validity, and I have consequently attempted to use them as a partial guide in laying out a simplified course of instruction in the methods of identifying crystals by means of the polarizing microscope. For example, the text is confined to one concept (the optical indicatrix) in its discussion of crystal optics. In this approach it pays homage to the still pertinent statements of F. E. Wright (1923, p. 780):

In crystal optics a number of surfaces of reference have been derived and are described in the text books to aid the student in visualizing the spacial geometrical relations involved; but if presented one after another in rapid succession, as is commonly the case, they tend to confuse and to bewilder him; he finds it difficult to distinguish between them and to realize the significance of each; and, as a result, may fail to obtain a clear idea of any part of the subject. The remedy is obvious, namely, select the particular surface of reference best suited to the purpose at hand and adhere to it strictly to the practical exclusion of the other surfaces. In petrographic microscope work the interest centers chiefly in the effects produced on plane polarized waves on transmission through a crystal plate; this means primarily a consideration of wave fronts, wave normals, and refractive indices; for this purpose the index ellipsoid and the surface derived from it, the index surface, furnish the most direct mode of presentation. To employ any but the simplest possible surface of reference would appear to add to the difficulty of forming a mental picture of the optical and crystallographical relations. Once the simplest surface has been thoroughly mastered by the student it is easy for him to add other surfaces and to perceive their significance. It is believed that by thus reducing to a minimum the geometrical conceptions required by the petrographer a real advance can be made in the presentation of the subject.

This text has been written specifically for the beginning student in optical crystallography. Description of advanced techniques or of numerous alternative techniques have been largely omitted in favor of a rather thorough but simple presentation of the basic techniques for determining the optical constants of crystals, using only a polarizing microscope and immersion media. Both the theory and the practical methods of optical crystallography have been accorded coverage in a proportion that, it is hoped, will best suit the needs of the student. The determination of the refractive indices of solids by the oil immersion method has been discussed in more detail than usual, following the suggestion of Emmons and Gates (1948) that more use be made of the colored dispersion fringes at grain boundaries. A set of determinative tables whereby over 1000 mineral specimens may be identified from their refractive index data is included in the Appendix. This organization and emphasis have been followed in the hope that the student in optical mineralogy will no longer need to consult three different types of books—one for theory, a second for practical methods, and a third for determinative tables—in order to attain an adequate grasp of the subject.

It is a pleasure to acknowledge my indebtedness to the teaching of Professor D. Jerome Fisher of the University of Chicago whose course both greatly challenged and stimulated the writer. Professor Horace Winchell of Yale University read the preliminary manuscript with great care and offered many suggestions that resulted in its improvement. Several of the writer's students, in particular Mr. Gerald V. Gibbs now of the Pennsylvania State University, read portions of the manuscript in its early stages and pointed out areas where the beginning student might have difficulty. Mr. George Desborough compiled the data for the determinative tables and Mr. Takashi Fujii aided in the proof reading.

All suggestions or corrections whereby the text may be made of greater service to the beginning student in optical crystallography will be greatly appreciated.

F. D. B.

Carbondale, Illinois
January, 1961

CONTENTS

ONE ⎤ LIGHT AND RELATED PHENOMENA

NOMENCLATURE

Gamma rays, x-rays, ultraviolet rays, visible light, infrared, and radio waves are all portions of the electromagnetic spectrum (Fig. 1–1). Each of these wave types travels, in a vacuum, at the common velocity (c) of 3×10^{18} Å* per second, the speed of light; each has a slightly different wavelength (λ) and frequency (f) from its nearest neighbor. Classified on the basis of wavelength in air, each part of the electromagnetic spectrum embraces a continuous range of wavelengths. Occasional overlaps in nomenclature exist; for example, rays ranging from 10 Å to 1.0 Å in wavelength are called gamma rays by some workers but x-rays by others.

Visible light represents a relatively limited band of wavelengths within the electromagnetic spectrum, ranging from 3900 to 7700 Å. By photochemical processes not yet understood, light of a particular wavelength within this range, if incident upon a normal human retina, produces a message that is interpreted within the brain as a particular color. The expanded scale of Fig. 1–1 denotes the wavelength limits set by Hardy and Perrin (1932, p. 16) for the seven distinctive "colors of the rainbow" recognized by Sir Isaac Newton. P. J. Bouma (1947, p. 16), however, eliminates Newton's indigo and places the wavelength limits of the color responses as follows:

Wavelength (Å)	3800–4360	4360–4950	4950–5660	5660–5890	5890–6270	6270–7800
Color sensation	violet	blue	green	yellow	orange	red

---

* One angstrom unit (symbol Å) equals 10^{-8} cm or 1/10 of a millimicron (symbol mμ).

Bouma's longer limits for the visible range, 3800 to 7800 Å (as compared with the oft-cited 3900 to 7700 Å), are probably inspired by the facts that (1) some human eyes can detect light of longer or shorter wavelengths than normal and (2) highly intense light sources may emit light at the extremities of the range of sufficient energy to stimulate a response in even a normal eye.

Each color of the visible spectrum, as may be noted in a rainbow, grades imperceptibly into its neighbor. Understandably, therefore, the preceding wavelength values delimiting the distinctive colors are somewhat arbitrary.

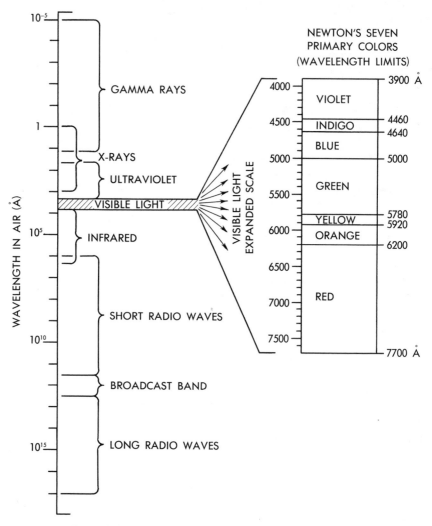

Fig. 1-1. The relation of visible light to the electromagnetic spectrum is shown on the left (modified from Sears and Zemansky, 1952). The expanded scale on the right illustrates the range in wavelengths assigned by Hardy and Perrin (1932) to Newton's seven principal colors.

One would have much difficulty, for example, in deciding whether light of 6250 Å wavelength was orange or red. As a matter of fact, Bouma's limits differ from the Hardy-Perrin limits (Fig. 1–1) with respect to this region.

If light of all wavelengths (from 3900 to 7700 Å) simultaneously strikes the human retina, the light is interpreted by the brain as "white light." Monochromatic light, on the other hand, refers to light with a much narrower range of wavelengths; the narrower the range, the more highly monochromatic it is. A sodium vapor lamp, for example, is a source of highly monochromatic light since it chiefly emits light of wavelengths 5890 and 5896 Å. A tungsten incandescent lamp, daylight from a north window, or direct sunlight, however, are polychromatic; that is, light energy is emitted at many different wavelengths (Fig. 1–2). Optical measurements

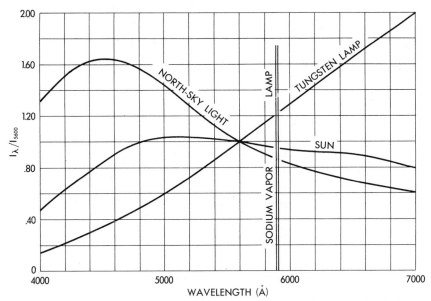

Fig. 1-2. The light energy emitted at various wavelengths, symbolized I_λ, as compared to that emitted at wavelength 5600 Å—that is, I_{5600}—for three polychromatic light sources: (1) the sun, (2) a typical gas-filled tungsten lamp (internal temperature 2600° C), and (3) north-sky light. If I_λ were plotted rather than I_λ/I_{5600}, the sun's curve would be far above the other two. The energy emitted by a suitably filtered sodium vapor lamp, chiefly light of wavelengths 5890 and 5896 Å, is indicated by two vertical lines. Technically, the heights of these two lines are infinite, not because I_{5890} or I_{5896} is so large, but rather because I_{5600} equals zero for this monochromatic source. (In part after W. D. Wright, 1958.)

of great accuracy require highly monochromatic light sources, and the sodium vapor lamp is commonly used. In routine work, however, either daylight from a north window or, more frequently, a tungsten lamp equipped with a blue "daylight" filter is used.

POLARIZED LIGHT

Light energy is generally considered to travel by means of a transverse wave motion in which the vibration of the particles is usually perpendicular to the direction of travel of the energy. As schematically shown on the left half of Fig. 1–3, ordinary (unpolarized) sodium light is considered to vibrate in numerous directions, all of which are at right angles to its ray path. As the ray of light energy travels from plane 1 to plane 11, its vibration directions may be imagined to trace out a three-dimensional figure resembling a series of canoes, end to end, with alternate canoes upside down. As shown, the ray's vibration directions within any plane perpendicular to its path may be represented by a semi-circle of radius equal to the vibration of the wave within that plane. Planes 1, 5, and 9 are exceptions since they are located at points on the ray path for which the vibration is nil.

Light whose vibrations are restricted to a single direction in space—for example, the ray between O' and O'' in Fig. 1–3—is said to be *plane polarized*. Those materials or devices that convert ordinary light to plane-polarized light are generally called polarizers. Light emergent from a polarizer has been made to vibrate parallel to one particular direction, which will be called, for simplicity, the "privileged direction" of the polarizer.* PP' represents the privileged direction of the polarizer located at plane 11 in Fig. 1–3. The *plane of vibration* of the light may be defined as the plane parallel to both the ray's path and vibration direction; in Fig. 1–3, therefore, it is the ruled plane parallel to the lines $O'O''$ and PP'.

Wavelength (λ) may be defined as the distance between two neighboring points experiencing vibrations of the same amount and direction, such

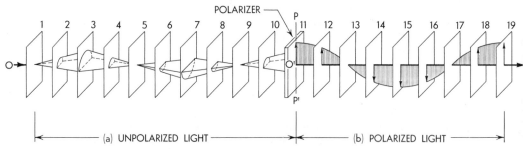

(a) UNPOLARIZED LIGHT   (b) POLARIZED LIGHT

Fig. 1-3. Unpolarized light traveling through imaginary planes 1 to 10 is polarized by a polarizer located at 11. After passing through the polarizer, the light vibrates exclusively parallel to PP', the privileged direction of the polarizer.

---

* Following several British physicists—for example, R. W. Ditchburn (1952, p. 369) and R. S. Longhurst (1957, p. 461)—the term "privileged direction" has been substituted for the more conventional term "vibration direction" in this text. Privileged direction refers to the vibration direction that light must observe while passing through a polarizer (or through an anisotropic crystal as defined on p. 6). A privileged direction exists for a polarizer even when it is not transmitting light whereas a vibration direction technically exists only during transmission. Substitution of the phrase "privileged direction of the polarizer" for "vibration direction of the polarizer" also prevents the student from visualizing the polarizer as being in motion.

points being said to be "in phase." Thus in Fig. 1–3 the wavelength for the unpolarized light equals the distance between construction planes 1 and 9 (or between planes 2 and 10); for the polarized light, the wavelength equals the distance between planes 11 and 19, the points at 11 and 19 being in phase. The amplitude of a wave is defined as the maximum vibrational displacement observed. The amplitudes are, for the unpolarized wave, the radii of the semicircles shown in planes 3 or 7 of Fig. 1–3; for the polarized wave, amplitude is represented by the vectors shown in planes 11, 15, or 19.

Although its amplitude appears larger than that of the unpolarized wave in Fig. 1–3, the polarized wave actually contains only one half of the total light energy of the unpolarized wave. Polarization of an ordinary light beam always produces a decrease in intensity. The reader will understand this more readily after study of future sections.

RELATIONSHIPS

The standard relationship between frequency (f), wavelength (λ), and velocity (c) for wave motion is

$$c = f\lambda \qquad \text{(Eq. 1–1)}$$

Hence, if two of the three quantities are known, the third can be calculated. For example, consider a wave of orange light whose wavelength in a vacuum equals precisely 6000 Å. Its speed in a vacuum, as for all light, is 3×10^{18} Å per second. Its frequency of vibration can thus be calculated as 5×10^{14} times per second. Similarly, the reader can easily calculate the frequency of sodium light (wavelength 5893 Å).

The frequency of a given beam of monochromatic light never changes, even if the light enters an entirely different material (its pulse rate, so to speak, remains constant). The wavelength and velocity of this same light, on the other hand, *do change* upon entrance into a different medium. Let the velocity, frequency, and wavelength of this light before and after it passes from medium A into medium B be indicated as c_A, f_A, and λ_A and as c_B, f_B, and λ_B, respectively. Because of the immutability of its frequency, f_A equals f_B. Coupling this fact and Eq. 1–1, it then follows that

$$\frac{c_A}{c_B} = \frac{\lambda_A}{\lambda_B} \qquad \text{(Eq. 1–2)}$$

that is, the wavelength of light entering a new medium changes in the same proportion as does its velocity. By way of example, if orange light of wavelength 6000 Å in a vacuum enters a medium in which its velocity is 1.5×10^{18} Å per second, its wavelength becomes 3000 Å.*

---

* An analogy may vivify Eq. 1–2. Assume that a pendulum, swinging at a constant rate, is placed within an elevator. As the elevator rises at a constant velocity, the pendulum bob approximately traces out a sine curve because its to and fro motion is coupled with the upward motion of the elevator. If the velocity of the elevator is doubled, the wavelength of the sine curve will also be doubled.

REFRACTIVE INDEX

The index of refraction (n) of a particular material may be defined as

$$n = \frac{c}{c_m} \qquad \text{(Eq. 1–3)}$$

where c and c_m symbolize the velocity of light in a vacuum and in the material, respectively. For most materials, c_m is less than c; consequently, refractive indices are generally greater than 1.0 in value. Air, through which light travels almost as fast as in a vacuum, has an index of refraction that may be assumed equal to 1.0 in most cases; actually, its index is approximately 1.0003. In general, the higher the density of a substance, the less rapidly light travels through it. High specific gravity and high refractive index are therefore related physical properties.

The wavelength of light entering a new medium changes inversely proportionally to its refractive index in the new medium. Thus

$$\frac{\lambda_A}{\lambda_B} = \frac{n_B}{n_A} \qquad \text{(Eq. 1–4)}$$

where n_A and n_B refer to the refractive indices of the two media involved. The derivation of Eq. 1–4 from Eqs. 1–2 and 1–3 is left to the reader.

ISOTROPIC AND ANISOTROPIC MEDIA

Those materials through which monochromatic light travels with the same speed, *regardless of its direction of vibration*, are called *isotropic* media. In addition to glass and crystals of the isometric system, a vacuum, all gases, and most liquids are isotropic with respect to light. Other materials, mainly the crystals of any nonisometric system, are *anisotropic* with respect to light; through them a light ray may travel with considerably different speeds for different directions of vibration within the crystal. Within isotropic media, the vibration direction of a light ray is always perpendicular to the ray path; within anisotropic media, the angle between vibration directions and ray path may be other than 90°.

QUESTIONS AND PROBLEMS

1. Light travels with a velocity of 2.25×10^{10} cm per second in water. What is the index of refraction of water? *Ans.:* 1.333.

2. Calculate the frequency of light whose wavelength is 4861 Å in a vacuum. *Ans.:* 6.1716×10^{14} cycles per second.

3. Assume the index of refraction of the clear, colorless jelly (vitreous humor) within the human eye to be 1.336. What is the wavelength of orange light (wavelength 6000 Å in a vacuum) while in this jelly? *Ans.:* 4491 Å.

4. What is the velocity of light while traveling in a glass whose index of refraction is (a) 1.5? (b) 1.9? *Ans.:* (a) 2×10^{18} Å per second; (b) 1.579×10^{18} Å per second.

TWO

LIGHT IN ISOTROPIC MEDIA

REFLECTION AND REFRACTION OF RAYS

A ray of light incident upon an interface between two isotropic media will generally give rise to a reflected ray (which never crosses the interface) and a refracted ray (which does). Fig. 2–1 illustrates a ray in air, IO, incident upon a glass block of refractive index 2.0. The plane of incidence ($STUV$) is defined as the plane that contains the incident ray (IO) and the line normal to the interface—(that is, NOM). It also contains the reflected ray (OL) and the refracted ray (OR) in every case. The angles of incidence (i), of reflection (l), and of refraction (r) are defined as the angles between the normal (NOM) and the incident, the reflected, and the refracted ray paths, respectively. In all cases the reflected ray path can be determined from the incident ray path since

$$i = l \qquad \text{(Eq. 2–1)}$$

Snell's Law

The relationship between the paths of the incident ray and of the refracted ray was determined by Snell in 1621 to be

$$n_i \sin i = n_r \sin r \qquad \text{(Eq. 2–2)}$$

where $n_i{}^*$ and $n_r{}^*$ refer to the refractive indices of the media in which the incident ray and the refracted rays travel, respectively. This relationship, known as Snell's law, permits calculation of the fourth value, given any three of them. In Fig. 2–1, for example, if $n_i = 1.0$, $n_r = 2.0$, and $i = 50°$, one can calculate that $r = 22\frac{1}{2}°$.

---

* In this and all following discussions, n_i will symbolize the index of refraction of the medium in which the incident ray travels and n_r will symbolize the index of the refracting medium.

Polarization by Reflection and Refraction

As illustrated in Fig. 2–1, the unpolarized ray *IO* gives rise to two partially polarized rays: (1) the reflected ray *OL* whose vibrations are chiefly perpendicular to the plane of incidence and (2) the refracted ray *OR*

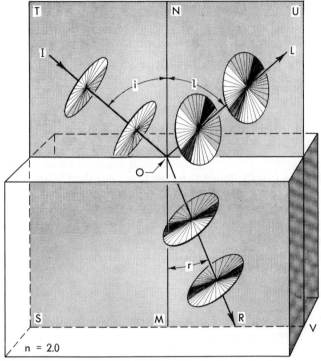

Fig. 2-1. An unpolarized ray, *IO*, incident in air upon the upper surface of a transparent glass block of index 2.0, forms a reflected ray, *OL*, and a refracted ray, *OR*. All the ray paths, as well as the angles of incidence (*i*), of reflection (*l*), and of refraction (*r*), lie within the plane of incidence, *STUV*.

whose vibrations are principally within the plane of incidence. Brewster in 1812 reported these two rays to attain a maximum degree of polarization when the angles of incidence and refraction were complementary—that is, when

$$\sin r = \cos i \qquad \text{(Eq. 2–3)}$$

Rewriting Eq. 2–2 to read

$$\frac{n_r}{n_i} = \frac{\sin i}{\sin r} \qquad \text{(Eq. 2–4)}$$

it is apparent from these two equations that

$$\frac{n_r}{n_i} = \tan i \qquad \text{(Eq. 2–5)}$$

The latter equation is known as Brewster's law. Thus, maximum polarization of the reflected and refracted rays may be produced by adjusting the angle of incidence until its tangent equals n_r/n_i. Complete polarization, however, is not attainable in this manner. If air is the medium containing the incident light ray, then Brewster's law becomes

$$n_r = \tan i \qquad\qquad \text{(Eq. 2-6)}$$

Critical Angle and Total Reflection

Snell's law (Eq. 2-2) may be rewritten

$$\sin r = \frac{n_i}{n_r} \sin i \qquad\qquad \text{(Eq. 2-7)}$$

Consequently, if the incident medium has a lesser index than the refracting medium, the fraction n_i/n_r must always be less than 1.0 in value. As a result, angle r must always be a smaller angle than angle i. For this case, therefore, true solutions of Snell's law exist for all possible values of angle i.

If the medium of incidence has a larger index than the refracting medium, the fraction n_i/n_r exceeds 1.0 in value. In this case, it is apparent (from Eq. 2-7) that angle r must exceed angle i. However, by its nature, angle r cannot exceed 90 degrees and its sine thus cannot exceed 1.0. Consequently, any instances wherein the quantity

$$\left(\frac{n_i}{n_r} \sin i\right)$$

exceeds 1.0 in value do not represent true solutions of Snell's law. Such anomalous results merely indicate that there was no refracted ray, the incident ray being entirely reflected. The value of angle i that makes the quantity

$$\left(\frac{n_i}{n_r} \sin i\right)$$

exactly equal to 1.0 is called the critical angle. For values of i less than the critical angle, total reflection of the incident ray does not occur.

The physical significance of the critical angle is illustrated in Fig. 2-2. The energy in each of the incident rays, OA, OB, and OC, is partitioned between a refracted ray (AA', BB', and CC', respectively) and a reflected ray (AO, BB'', and CC'', respectively). The relative intensities of the reflected and the refracted ray, roughly shown in Fig. 2-2 by different weights of lines, vary considerably according to the angle of incidence. For normal incidence—for example ray OA—the intensity of reflected ray AO is at a minimum.* As the angle of incidence increases,

---

* For normal incidence only, the relation between I_0 and I_l, the intensities of the incident and reflected rays, is

$$I_l = I_0 \frac{(n_r - n_i)^2}{(n_r + n_i)^2}$$

however, the reflected ray becomes increasingly more intense whereas the refracted ray becomes less so. Ultimately, when the angle of incidence reaches its critical value, the reflected ray possesses the full 100 percent of the incident ray's intensity. Thus light rays incident at angles equal to or greater than the critical angle—for example, rays OD and OE in Fig. 2-2—undergo total reflection, that is, none of the incident light energy is refracted across the interface.

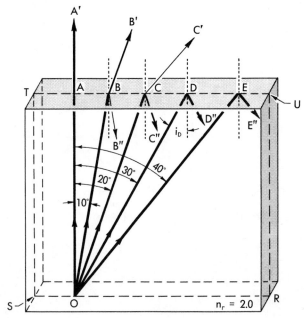

Fig. 2-2. A source of light at point O within a glass block gives rise to rays OA, OB, OC, OD, and OE (among others). The relative intensities of the associated incident, reflected, and refracted rays are approximately indicated by the use of lines of different weight. The vibrations are omitted. Ray OD is incident at the critical angle since for it, $i_D = 30° =$ arcsin n_r/n_i. ("Arcsin n_r/n_i" should be read "the angle whose sine equals n_r/n_i.")

Refraction of Light across Planar Surfaces

A light ray successively incident upon two parallel surfaces of a transparent body will always emerge parallel to the direction along which it entered. In Fig. 2–3A, for example, the equality between i_1 and r_2 is readily proven by successive application of Snell's law at the bottom and top interfaces.

A light ray successively incident upon two nonparallel planar faces of a transparent body—for example, an optical prism—will generally not emerge from the second plane along its original direction. For example, in Fig. 2–3B, where the glass prism is surrounded by a medium of lower refractive index, the exit ray is deflected toward the thick end (shaded) of the prism. However, in an oil of higher refractive index than the glass of the prism (Fig. 2–3C), the ray is deflected away from the thick end of the prism. If the solid and the surrounding medium have the same refractive index, light rays pass through the solid without deflection (Fig. 2–3D).

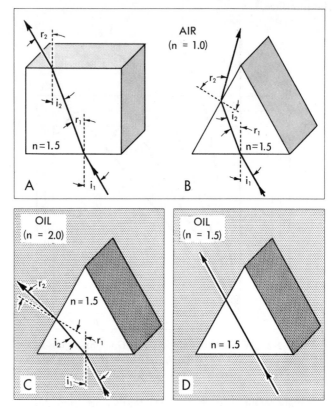

Fig. 2-3. Successive refraction of a light ray by two parallel interfaces (A) and by two nonparallel interfaces (B), (C), or (D). The front half of each glass solid has been removed to expose the plane of incidence. In (C) the glass prism is immersed in an oil of larger index than the glass; in (D) the glass and oil have identical indices.

DISPERSION

Cauchy's Equation

The separation of a ray or beam of white light into its component colors after the beam enters a second medium is known as dispersion. It occurs in media wherein the velocity, and therefore the index of refraction, varies for different wavelengths of light. The approximate relationship between the value of the wavelength (λ) and the material's refractive index for that wavelength (n_λ) is, according to Cauchy,

$$n_\lambda = A + \frac{B}{\lambda^2} + \frac{C}{\lambda^4} \cdots \qquad \text{(Eq. 2–8)}$$

Thus, if the corresponding indices of refraction (n_{λ_1}, n_{λ_2}, and n_{λ_3}) of a medium are known for at least three, preferably widely different, wavelengths of light (λ_1, λ_2, and λ_3), then each pair of values (for example, λ_1 and n_{λ_1}) may be successively substituted into Eq. 2–8 to yield three linear equations (whose only unknowns are the Cauchy constants A, B, and C). The actual values of A, B, and C can therefore be determined for the medium by simultaneous solution of these equations.

Dispersion of Sunlight

Assume that a glass prism has indices of refraction of 1.600 and 1.500 for light of wavelengths 3900 Å and 7700 Å, respectively. If, as in Fig. 2–4A, this prism is arranged with its vertical edge parallel to the slit, SS', only a very narrow beam of sunlight would impinge upon the prism. The same angle

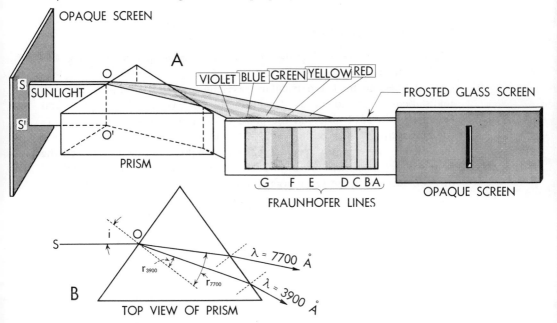

Fig. 2-4. (A) Prism dispersing sunlight from slit SS'. Dark lines A to G in the resulting spectrum are the Fraunhofer lines. Note that if the opaque screen is slid toward the left, its slit would isolate monochromatic light of a relatively narrow range in wavelength. (B) Top view of prism. Note that two different angles of refraction, r_{3900} and r_{7700}, result from the same angle of incidence, i.

of incidence, i, would pertain for SO, $S'O'$, and all the other rays of white light in this beam. However, the angles of refraction after the entry of each of these rays into the prism would be different for each wavelength of light. Thus, for solar ray SO (Fig. 2–4B), r_{7700} and r_{3900}, which are the angles of refraction within the prism for light of wavelengths 7700 Å and 3900 Å, respectively, may be calculated from Snell's law (Eq. 2–2) to be

$$\sin r_{7700} = \frac{\sin i}{1.500}$$

and

$$\sin r_{3900} = \frac{\sin i}{1.600}$$

Consequently, the angle r_{7700} exceeds angle r_{3900}, as is generally the case for all substances. The angles of refraction for all the intermediate wavelengths represented in the white light of the parent ray, SO, would be somewhere between r_{7700} and r_{3900} in value, depending on the particular wavelength. As a result, SO, upon entry into the prism, forms a fan of rays of colored light, each ray being composed of light whose wavelength differs slightly from that of its neighbor. $S'O'$, and all the parallel rays between it and ray SO, would be similarly dispersed into fans of colored rays vertically below the fan illustrated in Fig. 2–4B.

Fraunhofer Symbolism

The rainbow-colored wedge of light leaving the prism in Fig. 2–4 is called a spectrum. It can be regarded as consisting of a series of contiguous vertical lines, each of a particular wavelength of light, these component colors being referred to as "spectral colors." Wollaston in 1802 and Fraunhofer in 1814 observed the presence of numerous dark vertical lines (shown on the glass screen in Fig. 2–4A) in the solar spectrum, an indication that certain wavelengths are absent from sunlight. Fraunhofer denoted the more important of these dark lines by the letters A to G. The wavelengths corresponding to these Fraunhofer lines (Table 2–1) are of general interest

Table 2–1

WAVELENGTHS OF FRAUNHOFER LINES

Symbol of line	Wavelength (Å)
A	7594
B	6870
C	6563
$\left.\begin{array}{l} D_1 \\ D_2 \end{array}\right\} D$	$\left.\begin{array}{l} 5896 \\ 5890 \end{array}\right\} 5893$
E	5269
F	4861
G	4308

with respect to the elements present in the sun's atmosphere. The letter symbols are often used to refer to light of these particular wavelengths: the index of refraction of a substance for light of wavelength 5893 Å, for example, can be symbolized as n_D rather than n_{5893}. For refined optical work, the indices of refraction of a substance are generally determined for monochromatic light corresponding in wavelength to the C, D, and F values in Table 2–1. Precise measurements on crystals of NaCl indicate $n_C = 1.5407$, $n_D = 1.5443$, and $n_F = 1.5534$. If only one index is reported, it is generally n_D. This is probably because a filtered sodium vapor lamp furnishes the most intense, highly monochromatic source of light readily available to microscopists.

Coefficient of Dispersion

As already noted, the separation of a ray or beam of white light into its component colors is known as dispersion. As became apparent in Fig. 2–4, the amount of dispersion—that is, the degree of angular separation—between light rays of two different wavelengths depends on the difference between the two indices of refraction of the substance for these wavelengths. The difference in a substance's refractive indices for the F (4861 Å) and C (6563 Å) Fraunhofer lines—that is, the value $n_F - n_C$—is generally known as the *coefficient of dispersion;* it is commonly cited in crystallographic literature and sometimes serves as a criterion for identification of an unknown. The mineral mullite, for example, is occasionally distinguishable from sillimanite only by its higher coefficient of dispersion.

Dispersive Power

The dispersive power of a material is commonly stated as

$$\frac{n_F - n_C}{n_D - 1}$$

This value, although somewhat superior to the coefficient of dispersion as a measure of the ability of a substance to disperse white light, is less frequently cited in crystallographic literature. Occasionally, particularly for liquids, the dispersive power is stated as the reciprocal of the foregoing value—that is,

$$\frac{n_D - 1}{n_F - n_C}$$

LIGHT ABSORPTION AND COLOR TRANSMISSION

General and Specific Absorption

The intensity of a light beam decreases with passage through a material medium, as some of the light energy is converted to heat while in transit. This effect, called light absorption, is more pronounced in some media than in others. Many substances exhibit a "general absorption" of all the wavelengths of visible light (Fig. 2–5, glass C); others show a "specific or selective absorption" wherein particular wavelengths are more markedly absorbed (Fig. 2–5, glass B). Sunlit, stained-glass windows are examples

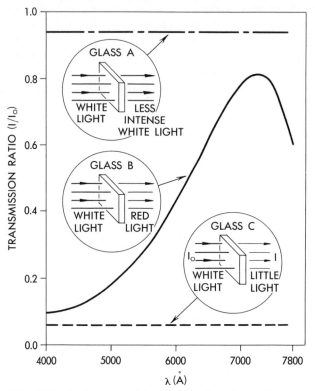

Fig. 2-5. Light transmission curves for plates of transparent colorless glass (A), red glass (B), and dark gray, almost opaque glass (C). I_o and I respectively signify the intensity of the light before entry and after passage through the glass.

of the color effects produced by selective absorption. The blue glass in the window, for example, preferentially absorbs the red wavelengths of the incident sunlight whereas the blue light of the sun's spectrum is transmitted, relatively unabsorbed, to the room interior. The red glass, on the other hand, preferentially absorbs the blue wavelengths to transmit red light (see Fig. 2–5, glass B).

Selective absorption is responsible for the transmission colors of certain minerals and crystals; the color transmitted is often a diagnostic property. In crystals the color is said to be *idiochromatic* if it is produced through selective absorption by the mineral itself, and *allochromatic* if it is the result of selective absorption (or scattering) of light by minute impurities dispersed through the mineral.

Relation to Thickness; Absorption Coefficients

The degree of absorption of light by a crystal is strongly dependent on its thickness. The relationship is expressed in Lambert's law,

$$\frac{I}{I_0} = e^{-kt} \qquad \text{(Eq. 2–9)}$$

where I_0 and I respectively signify the intensity of a beam of monochromatic light before and after passage through a thickness, t. The symbol e is the base of the natural logarithms and k is the *absorption coefficient* of the material in which the beam is traveling. Equation 2–9 may also be expressed

$$\ln \frac{I}{I_0} = -kt \qquad \text{(Eq. 2–10)}$$

where ln signifies a natural logarithm.

The value of the absorption coefficient may vary for light of different wavelengths within the same material. For example, if one were to plot for glass B in Fig. 2–5 the value of k (that is, $\ln \frac{I}{I_0} \div -t$) on the ordinate instead of I/I_0, it is obvious that the resultant curve would still retain the upward slope of the original. In other words, for glass B the absorption coefficient would steadily decrease for the longer wavelengths of light. The advantage of plotting the absorption coefficient k instead of the transmission ratio is that the former is independent of thickness.

The thickness of the crystal may even affect the eye's interpretation of the transmitted color. For example, if glass B were 10,000 times thicker than the plate whose transmission curve is shown in Fig. 2–5, its transmission curve would be a nearly horizontal line even below the curve for glass C; in this thickness, glass B would transmit little light for any wavelength. This illustrates why some crystals appear black and opaque for certain thicknesses but at their thin edges may transmit a particular color. On the other hand, for a plate of glass B which is only 10^{-4} times the thickness of the plate whose transmission curve is shown in Fig. 2–5, the transmission curve would be similar to that for colorless glass A. Thus some minerals that are lightly colored in large masses may appear to be essentially colorless in small grains or thicknesses.

Effect on Dispersion

If, for a particular range of wavelengths, a crystal undergoes strong selective absorption, the dispersion curve will not follow the Cauchy relationship (Eq. 2–8) in the nearby wavelengths. Instead, the dispersion curve will be like the solid-line curves in Fig. 2–6. Except for the immediate region of the absorption band, these curves are well approximated by Sellmeier's formula

$$n^2 = 1 + \frac{A\lambda^2}{\lambda^2 - \lambda_0^2} \qquad \text{(Eq. 2–11)}$$

where A is a constant and λ_0 represents the wavelength of maximum absorption. For crystals with more than one absorption band the formula becomes more complex. Sellmeier's formula is approximated by Cauchy's if one can assume that (1) there is only one absorption band and (2) it is located far to the long wavelength side of the wavelength region being investigated. In Fig. 2–6A, it is apparent that, nearing an absorption band, the dispersion curve slopes downward more strongly than indicated by Cauchy's equation. Sellmeier's formula thus yields the better fit in this region, although within the absorption band it too breaks down.

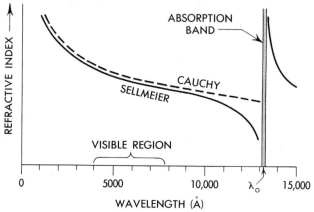

Fig. 2-6A. Dispersion curve (solid line) for a crystal showing "normal dispersion" in the visible region.

Since dispersion curves were first compiled for materials that were transparent throughout the visible spectrum, any absorption bands possessed by these materials were located, in general, well beyond the visible region. As a consequence, the concept of *normal dispersion* arose—namely, that the

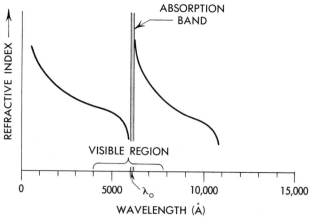

Fig. 2-6B. Dispersion curve for a crystal showing "anomalous" dispersion in the visible region.

refractive index decreased for the longer wavelengths throughout the visible range (Fig. 2–6A). Later work was extended to materials possessing absorption bands within or near the visible range. Their dispersion curves within the visible region were considered abnormal since refractive index sometimes increased for the longer wavelengths (as shown in Fig. 2-6B). Such materials were said to exhibit "anomalous dispersion."

QUESTIONS AND PROBLEMS

1. For the glass prism of Fig. 2–3B (interfacial angles of its sides are 60°), if $i_1 = 0°$, what is the value of r_2? *Ans.:* No correct value exists; the ray within the crystal undergoes total reflection when incident upon interface 2 (the upper left face) in Fig. 2–3B.

2. Assume that the glass prism of Fig. 2–3B is hollow, each side (interfacial angles again 60°) consisting of a thin plate of glass. The prism is filled with an unknown liquid, and sodium light is observed to enter the prism at interface 1 and leave from interface 2 at the angles $i_1 = 30°$ and $r_2 = 61°57.3'$. What is the refractive index of the liquid? *Ans.:* 1.400.

3. Calculate the values of constants A and B (neglect C), in Cauchy's equation for the mineral halite ($n_C = 1.5407$, $n_D = 1.5443$, and $n_F = 1.5534$). *Ans.:* $A = 1.5253 \pm .0005$; $B = 6.45$ to $6.73 \times 10^5 Å^2$.

4. Same question as 3, for fluorite ($n_C = 1.4325$, $n_D = 1.4338$, and $n_F = 1.4370$). *Ans.:* $A = 1.4270$; $B = 2.35$ to $2.37 \times 10^5 Å^2$.

5. The telescope of an alidade is located 5 feet above the liquid level but precisely over one corner of a 10-foot-square liquid-filled vat. The telescope's line of sight, when focused upon a point marked 10 feet below the liquid level on the diagonally opposite corner, makes an angle of 35°32.25′ with the horizontal. What is the refractive index of the liquid? *Ans.:* 1.4001.

6. (a) An electric lantern falls into a pond of water to a depth of 10 feet but continues to emit light in all directions. Assuming the pond water to be perfectly calm and transparent ($n = 1.333$), the owner, if in a canoe that forms no ripples, would detect light from it only within a radius of how many feet immediately above the water's surface? (b) How could he improve his chances of locating the lantern? *Ans.:* (a) 11.34 feet; (b) Place his head below water surface and look around.

7. What is the angular altitude above the horizon at which the sun appears to set if viewed by a fish within the ocean (assume $n = 1.33$ for ocean water)? *Ans.:* 41°25′.

THREE ___ $\textbf{L}$ENSES AND

THE COMPOUND MICROSCOPE

NOMENCLATURE AND TYPES OF LENSES

A lens may be defined as a body of glass (or other transparent isotropic substance) that is bounded by no less than one curved surface. The shape of lenses bounded by spherical surfaces (the simpler types) is dependent

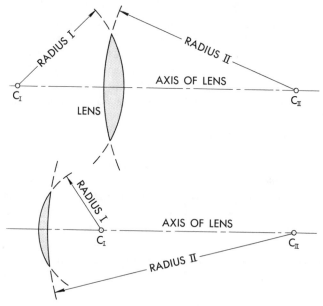

Fig. 3-1. Two examples of the relationship between the cross-sectional shape of a lens (front view circular in all cases) to: (1) the radii of curvature of its spherical surfaces and (2) the relative locations of C_I and C_{II}, their centers of curvature. The line passing through the centers of curvature and the center of the lens is called the axis of the lens.

upon the radii of curvature of these surfaces as well as upon the relative locations of their centers of curvature (Fig. 3–1). The *axis of a lens* is the line connecting its physical center with its centers of curvature. By definition a *thin lens* (not to be confused with the term "thin-edged lenses" discussed in a following paragraph) is one whose thickness is small as compared to its radii of curvature.

The six basic shapes of lenses resulting from different combinations of radii of curvature and relative locations of centers of curvature (Fig. 3–2) may be grouped into two types: (1) those that are thinner at the edges than at the center and (2) those that are not. The planar surfaces on lenses are considered to have an infinitely long radius of curvature.

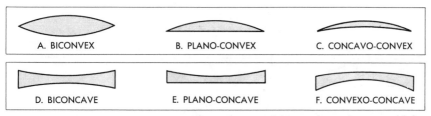

Fig. 3-2. Cross sections through different types of lenses (top views would be circles): Top (A, B, C), thin-edged (converging) lenses; bottom (D, E, F), thick-edged (diverging) lenses.

Thin-edged lenses, the most important lens elements of polarizing microscopes, converge all monochromatic light rays traveling parallel to the lens axis so that they intersect at a point called the *principal focus* (point *O* in Fig. 3–3A); thick-edged lenses, less important for our purposes, diverge

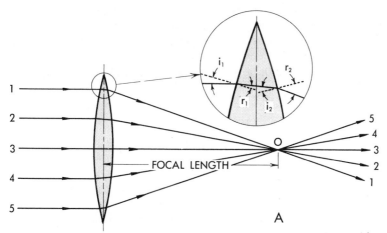

Fig. 3-3A. Focusing of light rays parallel to the lens axis by a thin-edged lens (in cross section). The detailed view shows how the refraction at the lens surfaces is in accordance with Snell's law.

such rays so that they appear to have emanated from a point, again called the *principal focus* (Fig. 3–3B). In each case the distance of the principal focus from the lens' center is called the *focal length* of the lens. Note that in Fig. 3–3A a *real* point source of light is created at *O*; that is, the rays truly cross at and therefore secondarily emanate from point *O*. However, in Fig. 3–3B, *O'* is a *virtual* point source; that is, it appears to the eye that the rays emanate from *O'*, but actually they do not.

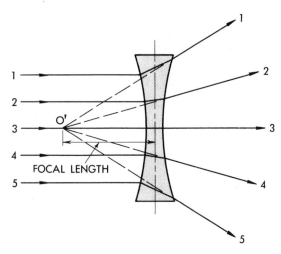

Fig. 3-3B. Focusing of rays by a thick-edged lens.

THIN-LENS FORMULA AND GRAPHIC SOLUTIONS

A thin-edged lens intercepting the light rays emitted from a point, *P*, on its axis will cause these rays to converge to a point (Fig. 3–4A), become parallel (Fig. 3–4B), or become less divergent (Fig. 3–4C); the effect depends upon whether the point source is located beyond, at, or within the focal length of the lens, respectively. Using *p* and *q* to symbolize the distance from the center of the lens to the point source and its refracted image, respectively, their relation to *f*, the focal length of the lens, is expressed for thin lenses as

$$\frac{1}{f} = \frac{1}{p} + \frac{1}{q}$$ (Eq. 3-1)

The values *f* and *p* are always positive in sign but *q* may occasionally be negative. A negative value of *q* indicates that the image of the point is a virtual one and is located on the same side of the lens as the original point (Fig. 3–4C). The distances *p* and *q* are called conjugate distances; similarly points *P* and *P'* (Fig. 3–4A, C) are termed conjugate foci.

The location of the lens-formed image of a point located slightly off the lens' axis—for example, the arrow tip in Fig. 3–5—can be determined graphically as a check on a solution of Eq. 3–1. Thus, of the infinite number of light rays emanating from the arrow tip, the paths of three of

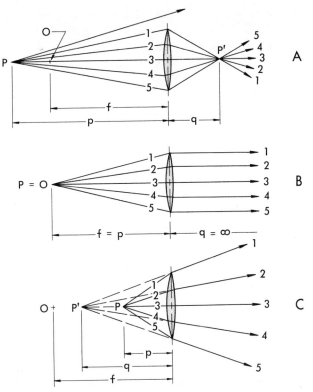

Fig. 3-4. Relationship between p (the distance of P, a point source of light, from the lens center), f (the focal length of the lens), and q (the distance from the lens center of P', the refracted image of the point source). (A) p exceeds f: P and its image, P', lie on opposite sides of the lens. After passage through the "crossroads" at P', ray 1 becomes the lowermost of the rays rather than the uppermost. (B) $p = f$: No image P' is formed. The rays are parallel after passage through the lens. (C) p is less than f: P' now lies on the same side of the lens as P. Rays 1 to 5, viewed after passage through the lens, appear to have emanated from a point at P'.

them are known even after they pass through the lens. Their intersection (real or apparent) will thus establish P', the point where *all* of the light rays from the arrow tip P either intersect (Fig. 3–5A) or appear to intersect (Fig. 3–5B). Specifically, the paths of these three rays (Fig. 3–5) are as follows: Ray 1 travels parallel to the lens axis before entry into the lens, then through O_F, the principal focus, after its refraction by the lens; ray 2 travels through the optical center of the lens and therefore emerges undeviated (planes tangent to the lens at this ray's points of entry and exit are parallel; thus this case is like that shown in Fig. 2–3A); ray 3 either passes through O_B, the back focus (Fig. 3–5A), or has a direction as if it had come from O_B (Fig. 3–5B) and therefore travels parallel to the lens axis after emerging from the lens. $R'P'$ (Fig. 3-5A) is a *real image* of PR whereas $P'R'$ (Fig. 3-5B) is a *virtual image*, only *seeming* to emit rays.

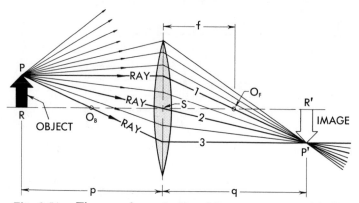

Fig. 3-5A. The use of rays 1, 2, and 3 to construct graphically the image $P'R'$ that a lens produces of an object PR located beyond its focal length, O_BS, or (see Fig. 3-5B, below) within O_BS. In Fig. 3-5B all rays being emitted from P, except for the three construction rays, have been omitted. In graphical constructions, rays are assumed to change direction at the central plane of the lens (vertical line at S) rather than, as actually occurs, at lens surfaces.

In Fig. 3–5 A and B, SPR and $SP'R'$ are similar triangles. Consequently

$$\frac{R'S}{RS} = \frac{P'R'}{PR} \qquad \text{(Eq. 3–2)}$$

However, $R'S$ and RS are, by definition, q and p, respectively. The ratio $P'R'/PR$ represents the ratio of image size to object size—that is, the magnification. Consequently, Eq. 3–2 can be rewritten

$$\frac{q}{p} = \text{magnification} \qquad \text{(Eq. 3–3)}$$

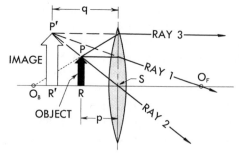

Fig. 3-5B. See caption for Fig. 3-5A.

THE COMPOUND MICROSCOPE

The compound microscope combines two converging lenses or systems of lenses, the objective and the eyepiece, each mounted in fixed positions in the opposite ends of a metal tube of length L (Fig. 3–6). Basically, the objective lens forms an enlarged real image (of the object being examined)

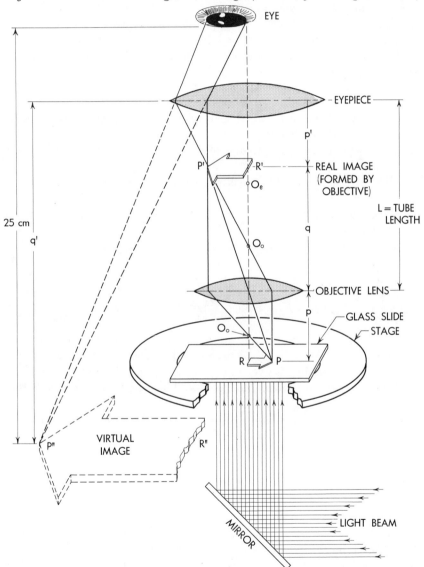

Fig. 3-6. The elements of a compound microscope as shown by a combined perspective view (stage, object RP, images $P'R'$ and $P''R''$) and cross-sectional view (rays, mirror, and lenses). The optical tube length, L, has been disproportionately shortened for illustrative purposes only.

at a point closer to the eyepiece than the eyepiece's focal length. Consequently, the already enlarged real image is seen through the eyepiece as a further enlarged virtual image.

The object to be viewed—RP in Fig. 3–6—is mounted on a glass slide and placed on the stage where it is illuminated by either daylight or lamplight (reflected upward by an adjustable mirror). When so illuminated, every point on the transparent arrow RP acts as a point source of light emitting an upward-directed hemispherical fan of rays in all directions. Considering one such point P and selecting from the hemispherical fan only the three special ray directions previously discussed, one notes that a real, inverted image of the object is formed at P'R' by the objective lens. Since it is a real image, every point on P'R' is again a point source of light emitting an upward-directed hemispherical fan of rays. The real image P'R', however, is formed closer to the eye lens than O_e, the principal focus of the eye lens; consequently, P'R', when viewed through the eyepiece, appears to be an enlarged virtual image, P''R'', located at a distance q' from the eyepiece.

The virtual image P''R'' will generally appear most distinct—that is, in best focus—if its apparent distance from the eye is 25 cm.* If the image P''R'' is not clearly seen, the microscope tube (containing the objective and eyepiece) may be racked up or down to alter the value of p and therefore of q, p', and q' until the virtual image is 25 cm away from the eye.

The magnifying power, M.P., of a compound microscope is approximately

$$\text{M. P.} = \frac{25 \text{ cm} \times L}{f_0 \times f_e} \qquad \text{(Eq. 3–4)}$$

where L, f_0, and f_e respectively represent the tube length, and the focal lengths of the objective and eyepiece in centimeters. Since for some eyes, 25 cm may not be the distance of distinct vision, a value other than 25 cm may need to be substituted into Eq. 3–4. Thus the same microscope may yield slightly different magnifications for different observers.

The *line of collimation*, or *axis*, of a microscope is the line coinciding with the axes of all its lenses. In Fig. 3–6, it is the line containing points R, O_0, O_e, and R'.

QUESTIONS AND PROBLEMS

1. A point of light is located along the axis of a biconvex lens at 3.2 inches distance from its center. If the focal length of the lens is 2.4 inches, determine the type of image and its distance from the lens' center. *Ans.:* Real image, 9.6 inches from center.

---

* The normal human eye is unable to see clearly any objects closer than 25 cm, the so-called distance of distinct vision.

2. A neon arrow is located on the axis of a biconvex lens ($f = 15$ cm) at a point 35 cm from the lens center. What is the magnification and type of image formed? *Ans.:* Real image, 0.75x.

3. Same question as 2, but the distance of arrow is only 10 cm. *Ans.:* Virtual image, 3x.

4. Two lenses are mounted in a vertical tube so that their axes coincide and their center-to-center distance is 16 cm. An object on this common axis is viewed through this double lens combination at a distance of 5 cm from the center of the nearest lens—that is, the objective lens. The focal length of this objective lens is 3 cm. The focal length of the lens nearer the eye—that is, the ocular—is 12 cm. What magnification is produced when the object is so viewed? *Ans.:* 5.15x.

FOUR — THE POLARIZING MICROSCOPE

The polarizing microscope is little more than a compound microscope into which two polarizers have been integrated (Fig. 4–1). One, located below the stage, is called either the lower nicol*, the lower polarizer, or simply the polarizer. The second, located above the stage, is called the upper nicol or analyzer. In most modern microscopes the polarizer transmits plane-polarized light vibrating in a north-south direction†, and the analyzer transmits only light (or that component thereof) that vibrates in an east-west direction. In some older models of microscopes and a relatively few modern ones, the privileged directions of polarizer and analyzer are reversed —that is, E–W and N–S, respectively.

If an object on the stage is viewed with the analyzer inserted, it is said to be viewed between *crossed nicols*. If the analyzer's privileged direction is made parallel to the polarizer's, the object is viewed with *parallel nicols*. If the analyzer is not inserted, the object is said to be observed with *plane light;* that is illuminated by plane-polarized light from the polarizer.

ELEMENTS AND THEIR FUNCTION

The Substage Mirror

The substage mirror of the microscope can usually be adjusted on either of its two axes so as to direct the most light upward along the axis

---

* The first polarizers used in microscopes were invented by William Nicol and were called Nicol prisms. The term "nicol" now refers to any of the various types of polarizing devices presently used in microscopes.

† The convention of referring to microscope directions as points of the compass perhaps arises from the youthful days of microscopy when an unobstructed north-facing window was considered to be a good light source. Today, consequently, an observer looking into a microscope is by custom assumed to face north, regardless of his actual orientation.

of the microscope. Its planar side is generally used for low and medium magnifications, its concave side for high magnifications.

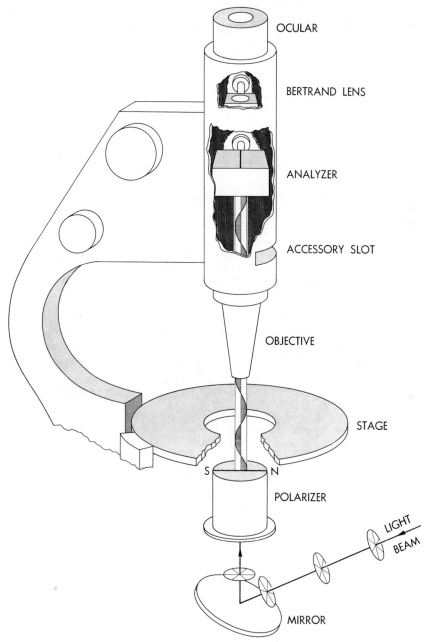

Fig. 4-1. Schematic diagram of the disposition of the more important parts of a polarizing microscope, mechanical details omitted.

The Substage Assembly

The substage assembly (Fig. 4–2) generally consists of a polarizer, a pair of condensing lenses, and an iris diaphragm. The entire assembly may be racked up or down to control its distance below the microscope stage and thus the illumination of the object.

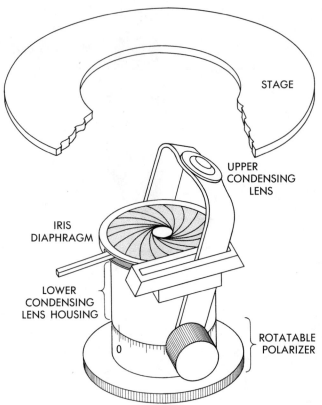

Fig. 4-2. Schematic diagram of the substage assembly of a representative polarizing microscope, mechanical details omitted.

Polarizer. As previously discussed, the polarizer in its normal position generally has a N–S privileged direction. In many models, however, the polarizer may be rotated on a vertical axis out of this position. If rotated back to normal operating position, it will make a slight "click" when the normal position is attained. In this position the index denoting the amount of rotation of the polarizer should indicate zero (see Fig. 4–2). If a microscope is being shared with another person, it is wise to make sure that the polarizer is in its zero position.

Lower Condensing Lens. This lens, located directly above the polarizer, cannot easily be removed from the path of the light. Its numerical aperture (see p. 32) should preferably be about equal to that of the microscope's medium-power objective—that is, about 0.25.

Iris Diaphragm. This is located above the lower condensing lens. Its aperture may be increased or decreased as desired by the turn of a lever.

Upper Condensing Lens. The uppermost element of the substage assembly is a powerful converging lens that in most microscopes can be swung into or out of the path of the light as desired. With this lens inserted, the light illuminating the object on the stage becomes more strongly convergent than if only the lower condensing lens is used. The numerical aperture when both condensing lenses are inserted should approximately equal the numerical aperture of the high-power objective with which it will normally be used. In some microscopes this condensing lens does not swing out but instead is slid up or down to increase or decrease its numerical aperture.

Rotatable Stage

The microscope's stage, onto which the microscope slides will be placed, should rotate freely and be calibrated so that its degrees of rotation can be determined on a vernier index. Angular measurements on microscopic preparations, which are very important in optical mineralogy, can then be readily made (although vernier accuracy is rarely needed). A pair of stage clips to hold the microscope slide firmly on the stage will decrease the possibilities for error in these measurements.

The Objectives

The objective lens of the microscope can be detached and exchanged for another by releasing the objective clutch, the design of which varies considerably among different makes of microscopes. Three achromatic objectives, respectively able to produce low (3.2x or 4x), medium (10x or 20x), or high (43x or 45x) initial magnifications* of an object, generally suffice for student work in optical mineralogy; the low-power objective provides a good overall view of a rock thin section whereas the medium and high powers show the section in more detail.

The characteristic properties of objectives with which a student should become acquainted are listed in Table 4–1 for three achromatic objectives commonly used in beginning work. As illustrated in Fig. 4–3, the angular aperture (A. A.) represents the angle between the most divergent rays that can enter the objective from a point on an object upon which the objective is focused. The angle equal to one half of the angular aperture is called u:

$$u = \frac{\text{A. A.}}{2}$$

---

* For example the magnification $P'R'/PR$ produced by the objective lens in Fig. 3–6.

For dry objectives, the numerical aperture (N. A.) equals sin u:

$$N.\,A. = \sin u \qquad \text{(Eq. 4–1)}$$

More broadly stated to apply to oil immersion lenses as well,

$$N.\,A. = n \sin u \qquad \text{(Eq. 4–2)}$$

Table 4–1
PROPERTIES OF THREE DRY* OBJECTIVES COMMONLY USED IN STUDENT WORK

Initial magnification	Angular aperture (A. A.)	Numerical aperture (N. A.)	Free working distance (F. W. D.) in mm	Depth of (clear) focus in mm
3.2x	14°	0.12	34.5	0.5
10x	29°	0.25	5.8	0.04
45x	116°	0.85	0.6	0.01

* Ordinary objectives, separated by an air space from the object being viewed, are called "dry" objectives. This is in contrast to the more powerfully magnifying oil-immersion objectives in which an immersion oil fills the gap between objective and object being viewed.

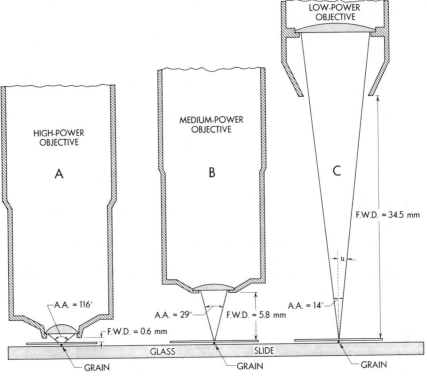

Fig. 4-3. Comparison of the free working distances (F. W. D.), angular apertures (A. A.), and one-half angular aperture (u) for the objectives described in Table 4-1.

where n equals the lowest index of refraction of any medium filling the space between the objective and the object being viewed. As may be noted in Table 4–1, the higher power objectives possesss larger angular and numerical apertures.

Of prime importance to the student is the concept of free working distance (F.W.D.), that is, the distance between the lowest part of the objective (in many cases a protective metal ring) and the top of the cover glass overlying the object in focus. As illustrated in Fig. 4–3, the free working distance is ordinarily very small for high-power objectives; therefore, rather than look through the eyepiece, the student should observe (with his eye at the level of the stage) the air space between objective and coverglass as the microscope is racked downward. When this air space is slightly less than the free working distance, he can look through the eyepiece and rack the microscope slowly *upward* until a sharp focus is secured. *Considerable damage to the object or objective may result if caution is not used during this operation.*

For some preparations the cover-glass thickness so far exceeds 0.16 to 0.18 mm, the preferred thickness, that the free working distance is less than zero for a high-powered objective and consequently the section cannot be brought into clear focus. Generally, however, when such difficulties arise for a petrographic thin section, the beginning student discovers that the thin section was placed upside down—that is, with cover glass down—on the stage of the microscope. If the cover glass is truly too thick, the student should switch to a lower-power objective.

An objective, inserted into a microscope, may produce fairly sharp images of points lying both above and below the particular point upon which it is precisely focused. *Depth of focus*, the distance between the upper and lower limit (Fig. 4–4), is an inverse function of numerical aperture, objectives

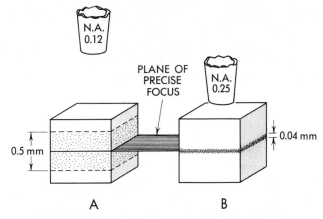

Fig. 4-4. Comparison of the depths of focus for (A) a low-power objective with N. A. 0.12 and (B) a medium-power objective with N. A. 0.25. The stippling represents the dust particles in a glass cube that are in fair focus if the objective is precisely focused upon the level of the ruled plane.

with large numerical apertures generally having small depths of focus and vice versa.

The importance of this feature in microscopy will become apparent if one focuses on the center of a glass cube throughout which dustlike inclusions are distributed, the microscope (10x eyepiece) being successively equipped with the three objectives described in Table 4–1. Using the 0.12 N.A. objective, all inclusions approximately 0.25 mm above or below the cube's center (Fig. 4–4A) would be in fairly sharp focus whereas with the 0.25 N.A. objective only the inclusions approximately 0.02 mm above or below this central point would be clearly visible (Fig. 4–4B). The high-power (0.85 N.A.) objective practically confines visibility to a plane through the point on which it is focused. Consequently, in studying crystals greater than 0.01 mm in thickness with a high-power objective, it is necessary to focus up or down to explore the crystal vertically for features that otherwise might be missed.

Accessory Slot

A slot, generally located in the lower end of the microscope tube (see Fig. 4–1), permits insertion of one of the three commonly used accessories (quartz wedge, full-wave plate, or quarter-wave plate) into the path of the light. The theory and utility of these three accessories will be discussed in a later section.

Analyzer

As already discussed, the polarizing element located above the accessory slot (Fig. 4–1) is called the analyzer and, as such, is readily withdrawable from the path of the light. In some models of microscopes, the privileged direction of the analyzer can be rotated as much as 90 degrees. For most operations, however, it is desirable that the privileged direction of the analyzer be at 90 degrees to that of the polarizer; to check this, insert the analyzer and observe whether, when nothing is on the microscope stage, complete extinction (blackness) exists, regardless of how strong the light source is.

Bertrand Lens

This lens, located above the analyzer (Fig. 4–1), may also be withdrawn from or inserted into the light path as desired. If it is inserted, the object in focus on the stage cannot be viewed; instead, under proper conditions, its interference figure (to be discussed later) will appear in the field of view. Microscopes whose Bertrand lens is equipped with an iris diaphragm permit observation of interference figures in smaller grains than is otherwise possible*.

---

* The Universal Eyepiece designed by F. E. Wright is particularly useful for observing interference figures in small grains but is generally not used in student work.

Ocular

A pair of negative (Huygenian-type) oculars, either 5x and 10x or 6x and 12x, usually suffices for routine microscopy. Generally speaking, oculars higher than 12x in power, unless used in conjunction with objectives of the very highest quality, do not yield very satisfactory images. For photomicrography, a positive (Ramsden-type) ocular of intermediate power

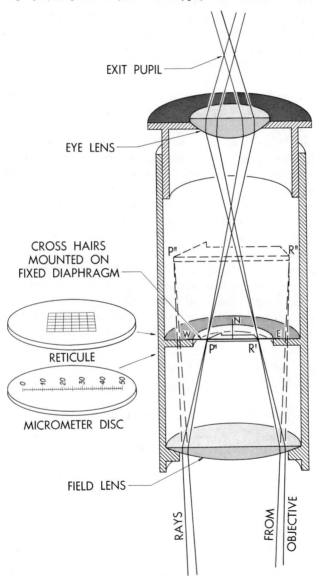

Fig. 4-5A. The components and ray paths within a negative (Huygenian) ocular. The dashed lines indicate the ray paths and real image location if the field lens were removed.

is also desirable. A student who finds it necessary to wear glasses during microscopy will generally prefer relatively low-power oculars since the location where one's eye sees the entire field of view—that is, the site marked "exit pupil" in Fig. 4–5A—is higher above the eye lens for low-power oculars. For some high-power oculars the exit pupil may be so close to the eye lens as to forestall observations by those wearing glasses.

A negative ocular (Fig. 4–5A) consists of two lenses. The lower lens (field lens) increases the convergency of the light rays from the objective so that the real image is formed at $P'R'$ rather than at the normal $P''R''$. A fixed diaphragm, which limits the field of view, is located precisely at the plane containing the real image $P'R'$. A pair of mutually perpendicular cross hairs (actually single dark threads, in some cases from a spider web) may be mounted on the fixed diaphragm so as to appear in the field of view as two mutually perpendicular black lines. Instead of these spider threads, a thin glass disc, upon whose lower surface is etched a pair of fine, mutually perpendicular lines, may be mounted on the fixed diaphragm to provide cross hairs. For quantitative linear measurements of the mineral grains being viewed, either a reticule or a micrometer disc (Fig. 4–5A) may be mounted in place of the cross hairs. Such oculars, called micrometer oculars or micrometer eyepieces, are available commercially. This fixed-scale type of micrometer ocular is adequate for most, if not all, optical or petrographic work. If greater accuracy is desired, a micrometer ocular with a moving scale should be obtained; its cost, however, is considerably greater. Either type of micrometer ocular must be calibrated for each objective with which it is to be used. This is done by focusing on a stage micrometer, which is simply a 3-by-1 inch glass slide whose central portion is engraved with a 2-mm-long line marked at intervals of 0.01 mm. When the scale of the stage micrometer is brought into focus (Fig. 4–5B, left scale), its divisions

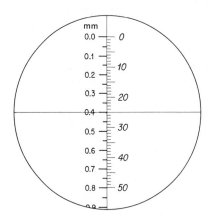

Fig. 4-5B. Calibration of the scale of a micrometer ocular (right side of N-S cross hair) by focusing on the scale of a stage micrometer (left side). The divisions indicating 0.01 mm have been omitted for the stage micrometer scale.

may be used to calibrate the divisions of the ocular (Fig. 4–5B, right scale). In Fig. 4–5B, for example, 50 ocular divisions embrace 0.8 mm of the stage micrometer scale. Thus each ocular division equals 0.016 mm for the objective-ocular combination whose field of view appears in Fig. 4–5B.

The upper part of the ocular (which houses the eye lens) slips sleevelike into the metal tube containing its field lens (Fig. 4–5A). This sleevelike motion permits the distance of the eye lens above the cross hairs to be adjusted so as to bring the cross hairs into sharpest focus for each observer. In turn, the entire assembly pictured in Fig. 4–5A fits into the microscope tube. During this insertion *under very gentle pressure*, the ocular assembly should be rotated until it slips into one or the other of the two slotted positions. Having slipped into one, it must be raised slightly before being rotated into the second position. For one of these positions the cross hairs will be observed to be in the standard NS-EW positions. For the second they will be at 45 degrees thereto.

Coarse and Fine Adjustments

The height of the objective above an object on the microscope stage must occasionally be varied in order to obtain a sharply focused image. This is generally done by two knurled knobs. One of these, known as the "coarse adjustment," racks the objective vertically upward or downward at a very noticeable rate; the second, known as the "fine adjustment," does so at a very gradual rate. The vertical movement produced by this fine-adjustment knob may generally be determined by counting the number of complete turns (or fractions thereof) of the knob by means of its attached calibrated drum. In some newer models the same knob controls both the coarse and fine motions, the fine-adjustment motion being obtained by reversing the direction of knob rotation; with this system, unfortunately, an inexperienced student is more likely to damage a thin section.

ADJUSTMENTS OF THE MICROSCOPE

Centering the Objective

An objective is "centered" when its lens axis (*bo* in Fig. 4–6A) coincides with the vertical axis about which the microscope stage rotates (*sa*). Fortunately the intersections of both axes with the microscope field of view are readily determinable. The outcrop of axis *bo* always coincides with the cross-hair intersection; the outcrop of *sa* can be located by viewing a dust speckled glass slide while rotating the microscope stage. One speck, at the point about which all the others revolve (like stars about the polar star), marks the outcrop of *sa* (point *a* in Fig. 4–6B). The cross-hair intersection can be brought to coincide with this point by turning the centering screws on the objective, by means of the two small centering wrenches provided. If the objective is now correctly centered, a speck seen at the cross-hair intersection will remain there during a 360-degree rotation of the stage. Each objective used with the microscope may require centering from time to time.

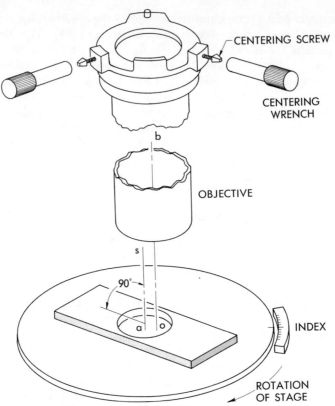

Fig. 4-6A. Uncentered objective. All parts of the microscope have been removed except for the stage and objective. The axis of the lens, *bo*, and the vertical axis about which the stage rotates, *sa*, do not coincide; therefore the objective is uncentered.

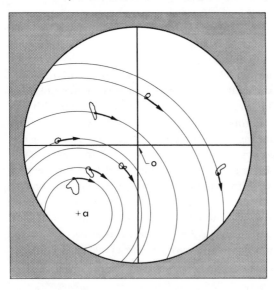

Fig. 4-6B. Field of view showing movement of particles on glass slide in Fig. 4-6A as the stage is rotated. The particles "orbit" around a point, *a*, which marks the outcrop of axis *sa* in the field.

Verification of Perpendicularity of Polarizers

Adjust the substage mirror and open the iris diaphragm until maximum brightness of the field of view is obtained. If the analyzer is now inserted, the field of view should be completely dark; if it is not dark, the polarizers are not mutually perpendicular. In that event, one should check as to whether the polarizer has been "clicked" (see p. 30) into its normal operating position and, in the case of microscopes with rotatable analyzers, whether the analyzer is in its zero position. With both polarizers in their zero positions, true perpendicularity between their privileged directions should exist, a situation commonly described as "crossed nicols."

Alignment of Cross Hairs to Nicols

During routine measurements, the cross hairs in the ocular should be parallel to the privileged directions of the nicols. This parallelism may be tested by observing, between crossed nicols, an oil mount (see p. 60) containing $PbCl_2$, $HgCl_2$, or natrolite needles. If the desired parallelism exists, the needles will appear darkest when aligned parallel to one of the cross hairs (Fig. 4–7). If parallelism does not exist, it is probably because (1) one of the polarizers is not in its zero position or (2) the ocular is not fitted into its proper slotted position.

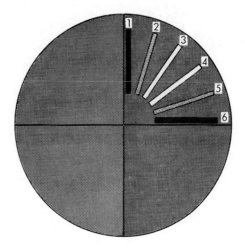

Fig. 4-7. Rotation of a needle of $PbCl_2$, $HgCl_2$, or natrolite successively into positions 1 to 6 between crossed nicols. The microscope used here has a correct cross-hair-nicol alignment since the needle is parallel to a cross hair when at maximum darkness (positions 1 and 6).

Cleaning Lenses

Clear images will be formed only if the glass surfaces of the microscope are free of fingerprints and of the film that sometimes collects on glass. To clean a lens, first brush it with a soft camel's hair brush to remove any abrasive dust particles, then wipe it with a lens tissue slightly moistened with xylene. Next polish the lens with a dry lens tissue. Since the ocular and objective lenses are handled relatively frequently, they are most likely to be smudged with fingerprints and should be checked for such if the microscope images are blurred. The glass of the slide on which the material being examined is mounted should also be free of oil droplets, smudges, and fingerprints. For cleaning slides, less expensive tissue may be used instead of lens tissue. For lenses, however, use only the abrasive-free lens tissue.

MODERN POLARIZING MICROSCOPES

The polarizing microscope is one of the more powerful tools at the disposition of the geologist, mineralogist, or chemist. The student should therefore become familiar with the name, function, and routine adjustment of each of the important working parts of his assigned microscope. To aid in this, labeled diagrams of the polarizing microscopes commonly used in optical mineralogy classes are included as Figs. 4–8 to 4–13.

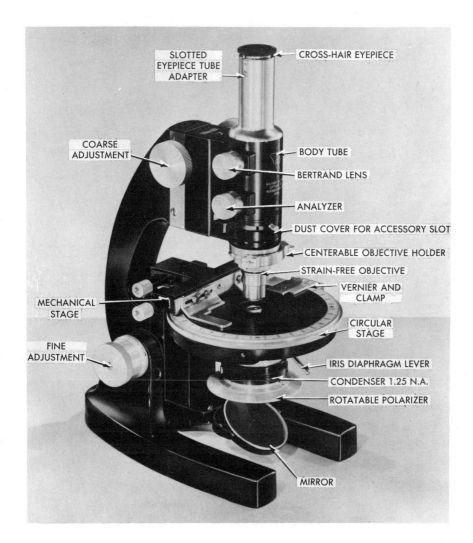

SLOTTED EYEPIECE TUBE ADAPTER

CROSS-HAIR EYEPIECE

COARSE ADJUSTMENT

BODY TUBE

BERTRAND LENS

ANALYZER

DUST COVER FOR ACCESSORY SLOT

CENTERABLE OBJECTIVE HOLDER

STRAIN-FREE OBJECTIVE

VERNIER AND CLAMP

MECHANICAL STAGE

CIRCULAR STAGE

FINE ADJUSTMENT

IRIS DIAPHRAGM LEVER

CONDENSER 1.25 N.A.

ROTATABLE POLARIZER

MIRROR

Fig. 4-8. Model LI polarizing microscope of the Bausch and Lomb Optical Company (Rochester 2, New York). The condenser may be slid vertically upward when light of increased convergence is desired.

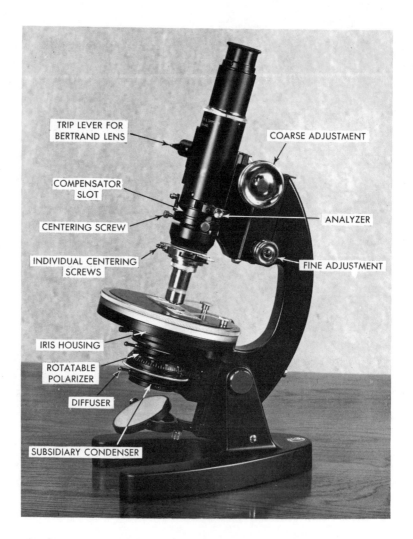

Fig. 4-9. Model M7110Q polarizing microscope of Cooke, Troughton, and Simms, Inc. (91 Waite Street, Malden 48, Massachusetts). Described in *Mineralogical Magazine* (1946, pp. 175-185).

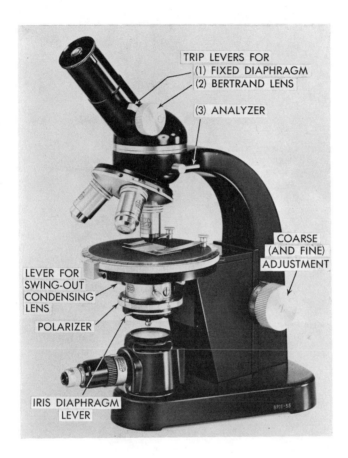

TRIP LEVERS FOR
(1) FIXED DIAPHRAGM
(2) BERTRAND LENS

(3) ANALYZER

COARSE
(AND FINE)
ADJUSTMENT

LEVER FOR
SWING-OUT
CONDENSING
LENS

POLARIZER

IRIS DIAPHRAGM
LEVER

Fig. 4-10. Model SM-Pol polarizing microscope of E. Leitz, Inc. (468 Park Avenue South, New York 16, New York).

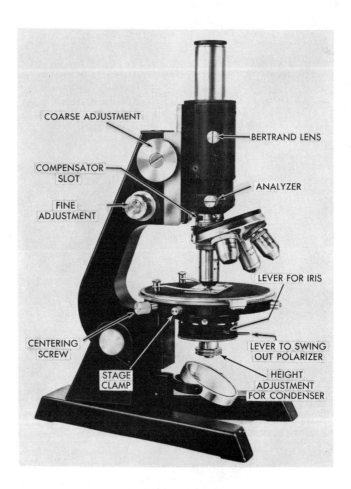

Fig. 4-11. Model RCP polarizing microscope of C. Reichert Optische Werke A. G. (Hernalser Hauptstrasse 219, Vienna XVII, Austria).

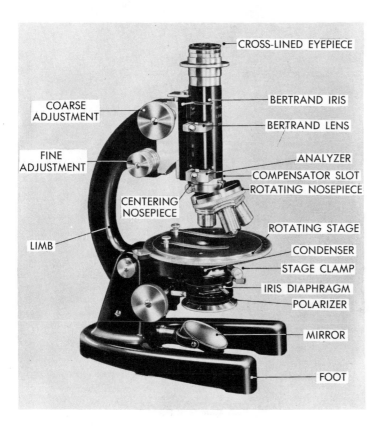

CROSS-LINED EYEPIECE

COARSE ADJUSTMENT

BERTRAND IRIS

BERTRAND LENS

FINE ADJUSTMENT

ANALYZER

COMPENSATOR SLOT

ROTATING NOSEPIECE

CENTERING NOSEPIECE

ROTATING STAGE

LIMB

CONDENSER

STAGE CLAMP

IRIS DIAPHRAGM

POLARIZER

MIRROR

FOOT

Fig. 4-12. Model P polarizing microscope of James Swift and Son, Limited (113-115A Camberwell Road, London, S. E. 5).

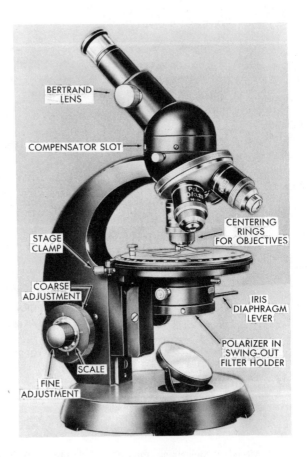

Fig. 4-13. Model KFT Standard Junior Pol polarizing microscope of Carl Zeiss, Inc. (485 Fifth Avenue, New York 17, New York).

FIVE

OPTICAL EXAMINATION
OF ISOTROPIC SUBSTANCES

As previously discussed, the refractive index of a substance differs for different wavelengths of light; hence, n_C, n_D, and n_F may all be measured in detailed studies. Commonly, however, the only index measured is n_D or n; n_D indicates that monochromatic sodium light was the light source used during the measurement whereas n indicates that white light was the illuminant during the measurement. Either value may suffice to identify an unknown material but n tends to be the less accurate. However, according to Emmons and Gates (1948, p. 612), if the colors into which the grain edges disperse the white light are properly interpreted, values of n that differ from n_D by less than $\pm$ 0.002 may be obtained.

The following discussions will therefore center on the measurement of n_D either directly or indirectly as n. It will be tacitly understood that n_C or n_F could be measured instead of n_D by substituting for the sodium light source either a suitably filtered hydrogen discharge tube or a monochromator set to emit wavelengths 6563 Å (C) or 4861 Å (F).

REFRACTIVE INDEX MEASUREMENT IN LIQUIDS

The Abbe refractometer has become the standard instrument for the measurement of n_D in liquids. Depending upon the model, the range of the refractive indices measurable may be 1.30 to 1.71 or 1.45 to 1.84 with an accuracy of $\pm$ 0.0002. A sodium vapor lamp or a monochromator set for 5893 Å is the usual illuminator. The less expensive Jelley microrefractometer may also be used to determine the indices of a liquid (within the range 1.33 to 1.92 and with an accuracy of $\pm$ 0.001). Both methods are rapid and convenient. Numerous alternative methods of refractive index measurement of liquids exist; in general, however, methods of increased accuracy also

47

require increased time. N. Bauer (Weissberger, 1949, Table X, p. 1238) summarizes the time required and maximum attainable accuracy for the various alternative methods. Fisher (1958) recommends use of a minimum amount of oil for increased accuracy during calibration of the Abbe refracto-meter with test plates of known indices.

REFRACTIVE INDEX MEASUREMENT IN SOLIDS

Duc de Chaulnes Method

This method permits measurement of the refractive index of transparent plates with only fair accuracy. It is included here briefly to point out the inverse relationship between a transparent plate's refractive index and its apparent thickness. The latter value represents the apparent depth of the plate's bottom surface below its top surface as viewed looking down through the plate. The three steps of the method are: (1) Focus a medium-power objective (N.A. 0.25) on the upper surface of a glass slide placed on the microscope stage (Fig. 5–1A); (2) carefully place the plate of unknown index

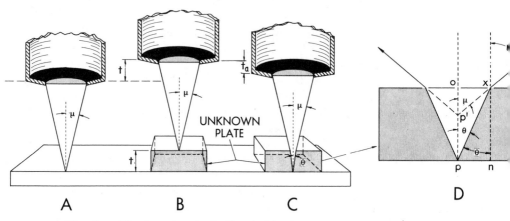

Fig. 5-1. The three steps in the Duc de Chaulnes method of measuring the refractive index of an unknown glass plate. (A) Focus on slide surface; (B) Focus on upper surface of unknown; (C) Focus on slide surface *through* unknown. In (D), an enlarged view of a portion of (C), the rays of light in air (xz, etc.) appear to have emanated from point p' although they actually originated from point p. Thus op' appears to be the thickness of the plate as measured by focusing through it. The angles θ and u have been exaggerated for illustrative purposes.

on this glass slide and then rack upward (using the fine-adjustment drum only) until the upper surface of the unknown plate is in sharp focus (Fig. 5–1B)—the difference between this second reading on the fine-adjustment drum and that obtained in step (1) indicates t, the true thickness of the unknown plate; (3) Now carefully focus downward through the unknown plate until the upper surface of the supporting glass slide is once more in sharp focus (Fig. 5–1C)—the difference between this fine-drum reading and that obtained in step 1 is a measure of the apparent thickness, t_a, of the unknown plate.

The Duc de Chaulnes equation for the unknown plate's refractive index is then

$$n = \frac{t}{t_a} \qquad \text{(Eq. 5–1)}$$

where t and t_a respectively signify true and apparent thickness. The differential drum readings obtained in steps 2 and 3 above are respectively proportional to t and t_a and may be substituted directly into Eq. 5–1 to obtain the refractive index of the unknown plate.

Derivation of Eq. 5–1, following Fig. 5–1D, is as follows:

$$\tan u = \frac{ox}{op'}, \text{ and } \tan \theta = \frac{ox}{op}$$

but op' and op respectively equal t_a and t; thus

$$\frac{\tan u}{\tan \theta} = \frac{op}{op'} = \frac{t}{t_a}$$

However, since rays px and xz obey Snell's law,

$$n = \frac{\sin u}{\sin \theta}$$

For small angles, the ratio of their sines approximately equals that of their tangents; consequently

$$n = \frac{\sin u}{\sin \theta} \cong \frac{\tan u}{\tan \theta} = \frac{t}{t_a}$$

Immersion Methods

One of the most convenient methods of measuring the refractive index of a transparent solid is by immersing fragments of it in a series of liquids of known refractive index. Criteria, which will be discussed in the following sections, can then be applied, to determine whether the refractive index of the unknown fragments is within 0.002 or 0.003 of that of the surrounding oil. The immersion liquids used should at least span the refractive index range between 1.430 and 1.740 at intervals of 0.004 or 0.005. Such sets of immersion media are obtainable commercially or can be prepared in the laboratory.

Relief. The depth of shadows along a grain's borders—that is, the degree of relief—grossly indicates how close the value of the grain's refractive index is to that of the oil. In a close match, the grain shows little or no relief if viewed in the oil; that is, it is almost completely invisible (Fig. 5–2A). To the extent that the indices of the oil and grain differ, however, the relief increases from low through moderate, high, and (ultimately) extremely high. Extremely high relief (Fig. 5–2B) is characterized by the presence of heavy shadows on the grain surfaces. The shadow effects that cause the grain to stand out in relief in the oil may arise because (1) the grain's index is less than that of the oil (negative relief) or (2) its index is greater than that of the oil (positive relief). Unfortunately, positive and negative relief cannot be distinguished through routine observation.

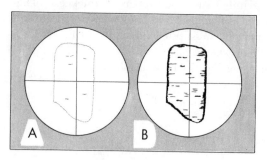

Fig. 5-2. Low relief (A) and high relief (B) of a grain immersed in oil.

The Becke Line Method*: A grain in oil, viewed with the microscope objective focused slightly above the position of sharpest focus, will usually display two thin lines (one dark and one bright) concentric with its border. The brighter of these is always closest to the material having the higher refractive index (Fig. 5–3, F_2F_2 fields of view) and, moreover, *always moves* (as shown by hollow arrows in F_2F_2 fields) *toward the medium having the higher refractive index, if viewed as the microscope is racked steadily upward above correct focus*—that is, from a focus upon plane F_1F_1 to F_2F_2. This line, which represents a concentration of light because of refraction or reflection, or both, at the grain-oil boundaries, is called simply the Becke line in most works. In this text, however, it will be called the bright Becke line to distinguish it on occasion from its dark companion, which, although usually nameless, will here be called the dark Becke line. The Becke lines become particularly obvious as the substage iris diaphragm is closed down.

The Becke lines are generally attributed to (1) refraction at the lenslike edge of the grain or (2) total reflection at the grain-oil boundaries, or both. Their explanation by refraction presumes that a grain whose index is greater than the oil's acts as a converging lens (Fig. 5–3A) whereas a grain whose index is lower acts as a diverging lens (Fig. 5–3B). Consequently, the bright Becke line, observed as the microscope is racked upward from a sharp focus at F_1F_1 to F_2F_2, appears to move toward the grain (Fig. 5–3A) or toward the oil (Fig. 5–3B), whichever has the higher index. Total reflection at near-vertical grain-oil boundaries (Fig. 5–4) similarly deflects the near-

---

* Sometimes called the method of central illumination.

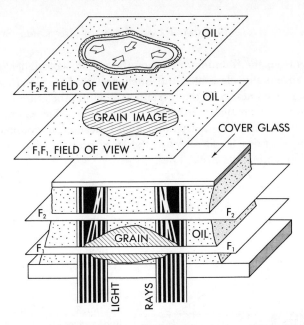

A. GRAIN INDEX GREATER THAN OIL'S

Fig. 5-3. The disposition and movement of Becke lines. Field of view F_1F_1 is that observed if the microscope is focused upon plane F_1F_1 to produce the sharpest grain image; here the Becke lines are not too obvious. However, if observed while the microscope is racked upward toward a focus on plane F_2F_2, the Becke lines become increasingly apparent, the brighter line moving (as indicated by the hollow arrows in field of view F_2F_2) toward the medium having the greater refractive index. The grain cross sections indicate the dispersionally produced "crowns" of darkness and of brightness (ray concentrations) that occur above grain-oil boundaries and give rise to the Becke line effects. The dark and bright crowns change position according to whether the grain has a greater index (A) or a lesser index (B) than the oil.

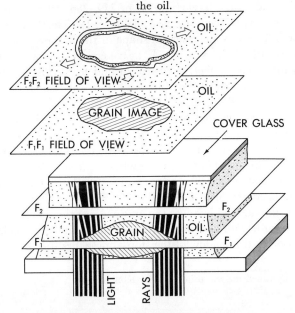

B. GRAIN INDEX LESS THAN OIL'S

parallel rays from the illuminator toward the medium of higher index and thus offers an equally plausible (and often preferred) explanation for the origin of the Becke lines. Note that certain rays within the grain are incident at angles equal or greater than their critical angle—for example, ray *a* in Fig. 5–4. Both effects, refraction and total reflection, undoubtedly occur at the edges of grains in oil and are therefore probably jointly involved in the development of the Becke lines.

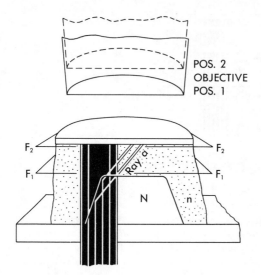

Fig. 5-4. The role of total reflection at a near-vertical boundary in producing the Becke line effect. As the objective is raised upward from its focus on plane F_1F_1 to a focus on plane F_2F_2 (that is, from position 1 to position 2), the Becke line moves toward the grain's center since, in this case, the grain has a larger index (N) than the oil (n). For simplicity, light rays on the right half of the grain and to the rear are omitted.

The sensitivity of the Becke line method is increased if the substage iris diaphragm is sufficiently closed so as to make the Becke lines distinct. Many authorities prefer the high-power (about 45x) objective for observing the Becke line effect, although Saylor (1935, p. 279, 286) cites a 10x objective as permitting increased accuracy (see Table 5–1). The sensitivity of the method is also affected by the size and shape of the fragments as well as by the intensity of the light source. A near-vertical crystal-oil boundary generally permits greatest sensitivity. According to Saylor (p. 294), more reliable results are obtained if the observed crystal-oil boundary is oriented parallel to the privileged direction of the polarizer.

Oblique Illumination Method.* Used with monochromatic light, this method permits detection, according to Wright (1913, p. 76), of differences as small as 0.001 between the refractive index of a grain and that of its surrounding oil. Thus, utilizing white light and a set of immersion media calibrated in steps of 0.005 or less, a grain's index of refraction can be measured to an accuracy of $\pm$ 0.003.

In essence, the method involves the gradual insertion of an opaque stop into the optical light path and observation of the grain in the field of view of the microscope as the shadow of the stop approaches it. Usually, a grain

---

* Occasionally called the Schroeder van der Kolk method.

whose refractive index is higher than that of the oil will become shadowed on the edges nearest the approaching shadow but bright on the opposite border (Fig. 5–5A); if the grain's index is lower than that of the oil, this will be reversed (Fig. 5–5B).

Fig. 5-5. Appearance under oblique illumination (shadow of opaque stop encroaching from left side of field of view) of a grain whose index is greater (A) or less (B) than that of the surrounding oil. For different microscopes and different locations of the opaque stop, the effect may be reversed.

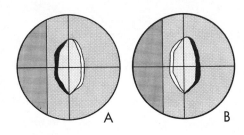

For certain microscopes (and for different locations of insertion of the opaque stop into the optical light path), "shadowed toward shadow" (Fig. 5–5A) may indicate the grain's index to be lower than that of the oil rather than greater. For such microscopes, "shadowed away from shadow" (Fig. 5–5B) will indicate the grain to have a higher refractive index than the oil. Consequently, for any microscope it is wise to check the significance of "shadowed toward shadow" and "shadowed away from shadow" by means of a grain of known index immersed in an oil of known index or, more simply, by a Becke line test on a grain already examined by oblique illumination.

Oblique illumination may be produced by nine or more ways, of which the following are most convenient: (1) slow insertion of a cardboard (or one's finger) into the light path below the polarizer (Fig. 5–6*a*); (2) partial insertion of an accessory into the accessory slot until its metal frame intercepts the light (Fig. 5–6*b*); or (3) partial insertion of the analyzer until its frame intercepts the light (Fig. 5–6*c*). Saylor (1935) suggests that a low-power objective increases the sensitivity of the method. The advanced worker may wish to use Saylor's "double diaphragm" method of oblique illumination, which significantly increases the accuracy of the technique.

Comparison of Methods. According to Saylor (1935, p. 285) the Becke line method is more sensitive than the ordinary oblique illumination method (but less so than his double diaphragm modification of oblique illumination). Saylor's results (Table 5–1) indicate that the medium-power objective (10x, N.A. 0.25), which he apparently terms a "low-power objective," yields the greatest accuracy in refractive index determination, being the best compromise between increased magnification and lowered aperture.

An excellent procedure for determining the refractive index of an unknown is to use the oblique illumination technique in the preliminary oil mounts, and the Becke line method for the final determinations when grain and oil are close in index. Here the advantages of oblique illumination

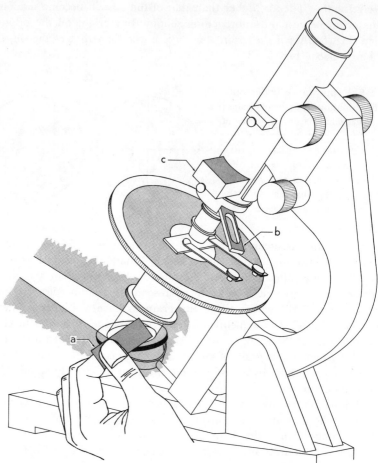

Fig. 5-6. Convenient production of oblique illumination by partly blocking light with cardboard at *a*, with metal frame of accessory at *b*, or with frame of (partly inserted) analyzer at *c*.

Table 5–1

(AFTER SAYLOR, 1935, FIG. 6)

Objective	Error in measuring refractive index of glasswool		
	by Becke line	By oblique illumination	
		ordinary	double diaphragm
10x, N. A. 0.25	± 0.0005	± 0.0005	< ± .0005
20x, N. A. 0.40	± 0.0009	± 0.0010	± .0005
45x, N. A. 0.85	± 0.0012	> ± 0.0020	± .0006

(rapidity and almost simultaneous testing of all grains within the field of view) are combined with that of the Becke line (higher accuracy than ordinary oblique illumination).

Dispersion Colors. If n_D of the grain is close in value to n_D of the surrounding oil, the bright and dark Becke lines (as well as the bright and shadowed areas produced by oblique illumination) will be observed to possess distinctive colorations if white light is the illuminant. Evidently, the white light is dispersed at the grain boundary into its spectrum (much as the prism dispersed the sun's rays; see p. 13). Close observation of these dispersion colors helps to determine whether n_D of the grain is greater than, equal to, or less than n_D of the oil. Consequently, n_D of the mineral can be estimated with accuracy better than $\pm$ 0.002 (Emmons and Gates, 1948, p. 612) even though white light rather than sodium light is the source.

As already discussed, the refractive index of an isotropic material differs for different wavelengths of light. Dispersion curves illustrating this for a typical grain and oil are shown in Fig. 5–7A. Such dispersion curves,

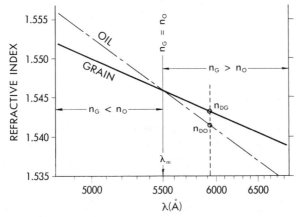

Fig. 5-7A. Dispersion curves (on Hartmann paper) of a typical oil and grain whose indices (n_o and n_g, respectively) are equal for light of wavelength λ_m (5500 Å). Note that, as is commonly the case, the slope of the oil curve is greater than that of the grain.

when plotted on Hartmann dispersion paper as in Fig. 5–7A, are usually straight lines sloping downward for the longer wavelengths of light; whereas on graph paper with an ordinary linear wavelength scale, they plot as curves. Significantly, the slope of these dispersion lines is usually less steep for a crystal than for an oil of similar index. Consequently, the dispersion lines representing an oil and a grain can never entirely coincide; at best they only can intersect at one point. This point marks λ_m, the wavelength of light for which both the oil and mineral have precisely the same index. In Fig. 5–7A, for example, λ_m equals 5500 Å and the refractive index of oil and crystal for light of this wavelength is a bit over 1.546.

Consider next, grains of this mineral mounted in this oil. If, as is commonly the case, the crystal is thinner at its edge than at its center, any white light incident at this edge will be dispersed into a spectrum (Fig. 5–7B). For light of wavelengths longer than λ_m, the grain's refractive

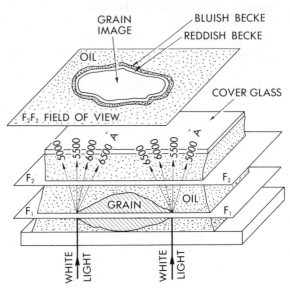

Fig. 5-7B. The dispersion of white light at the edges of the grain in the oil of Fig. 5-7A. Field of view F_2F_2 is that seen with the microscope no longer focused for sharpest grain detail on plane F_1F_1 but rather on plane F_2F_2. The angles between the rays of light for the different wavelengths are exaggerated.

index is higher than the oil's, thus the grain acts as a miniature condensing lens, serving to converge these rays. For light of wavelengths shorter than λ_m, the grain's refractive index is less than the oil's, thus the grain acts as a diverging lens for these shorter wavelengths. Light of wavelength λ_m, on the other hand, passes through oil and grain without deflection. As a result of these dispersional effects, the Becke lines rimming the grain are now definitely colored—although the novice may at first have difficulty in seeing these colors. Consequently, if observed as the microscope is being racked upward from F_1 (the position of sharpest focus) to F_2 (Fig. 5–7B), a red or reddish-orange Becke line is observed to move toward the grain's center whereas a violet blue or blue Becke line will move outward from the grain-edge into the oil. As may be deduced from Fig. 5–7A, the reddish Becke line is composed of wavelengths longer than 5500 Å, the bluish Becke line of wavelengths shorter than 5500 Å. A general rule thus becomes apparent: A *colored Becke line*, viewed as the microscope is focused upward, *moves toward the medium having the higher refractive index for the light wavelengths composing it.*

Interpretation of Becke Lines. In the process of refractive index determination by the immersion method, the grain may be placed in an oil whose refractive index differs so greatly from the grain that the dispersion lines do not intersect in the visible range (Fig. 5–8A). Consequently, whether in oil I or oil II, there would be no color observed in the Becke lines at this grain's edges. When in oil I, a very intense white Becke line would enter the oil as the microscope was racked up. For the grain in oil II, the results would be reversed, the bright white Becke line entering the grain. Lack of color in the Becke line thus indicates that the dispersion curves of grain and oil do not intersect within the visible spectrum.

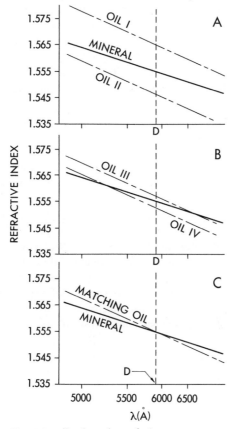

Fig. 5-8. Explanation of the movements and colors of the Becke lines according to whether (and where) the dispersion curves of the grain and oil intersect within the visible spectrum.

Consider grains of this mineral to be respectively immersed in two oils (III and IV in Fig. 5–8B) whose dispersion lines intersect that of the grain, but only at the opposite extremities of the visible spectrum. The Becke lines will now tend to develop slight colorations, although the bright Becke line will be still much more intense than the dark one. For the grain in oil III, a bright bluish-white Becke line will enter the oil as the microscope is racked up; simultaneously a dark reddish Becke line will enter the grain. In oil IV, on the other hand, a bright, yellowish-white Becke line will enter the grain whereas a less intense, dark violet Becke line enters the oil. In either case the bright Becke line enters the medium whose refractive index is higher for the majority of wavelengths of light. The particular Becke lines indicate, by their unequal intensities, that the dispersion lines of grain and oil intersect only at the outer fringe of the visible spectrum.

The ultimate goal of the immersion method is to immerse the grain in an oil whose refractive index (for sodium light) precisely equals that of the grain (Fig. 5–8C). *If sodium light is the illuminant,* the criterion for an exact match is the total disappearance of the Becke line, the grain (if uncolored and transparent) being completely invisible within the oil. *If white light is the illuminant,* the criteria for an exact match are (1) the approximate equality in intensity of the two Becke lines and (2) the color of these Becke lines.

According to Wright (1913, p. 75–76), equality in intensity of the two Becke lines signifies that "the refractive index of liquid and mineral are equal for a wavelength of about 5500 Å and the error of the determination of the refractive index n_D by this method is then not over $\pm$ .002... ." Wright says, however, that this probable error is noticeably larger when strongly dispersive liquids, such as methylene iodide (now usually called diiodomethane), are used. Actually, therefore, when refractive indices of the oil and mineral match for wavelength 5893 Å, the Becke line entering the oil should be slightly more intense than that entering the grain. For such a match, the Becke line entering the grain should be an orange-tinged yellow whereas that entering the oil should be a green-tinged blue.

If the set of immersion oils used consists of binary mixtures between mineral oil (1.470), α-monochloronaphthalene (1.633), and diiodomethane (1.739)—or oils of similar dispersion—the colors observed in the Becke lines (or in the oblique illumination shadows) are an important clue as to the difference in index, for sodium light, between the grain and the immersion oil (Fig. 5–9). These color interpretations presume the source of illumination to be a white tungsten bulb equipped with a blue "daylight" filter. The practical use of Fig. 5–9 is obvious. For example, a white Becke line is observed to enter an unknown mineral grain that is immersed in an oil of index 1.700. Fig. 5–9 indicates that the grain's index probably exceeds the oil's by 0.033 or more. The oil for the next mount should then be at least of index 1.733. The boundaries between the Becke line colors drawn in Fig. 5–9 are, of course, only approximate. Fig. 5–9 holds only to the extent to which the dispersion, $n_F - n_C$, of the mineral being examined approaches in value 0.015, the assumed average upon which the figure was based.

Effect of Temperature. The refractive indices of immersion oils decrease rather readily as temperature is increased, whereas those of solids are little affected. An instructive experiment is to place a grain in an oil whose index is sufficiently high so that its dispersion curve does not quite intersect that of the grain at room temperature (Fig. 5–10). Next gently warm the oil mount on a hot plate (do not boil) until a temperature such as T_5 has been attained; the oil's dispersion curve will have shifted translationally (from T_1 to T_5 in Fig. 5–10) whereas the grain's curve will

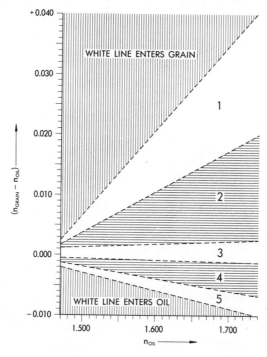

Fig. 5-9. Approximate relationship of the Becke line colors and movements to the refractive index difference (for sodium light) between grain and immersion oil (microscope racked upward). Area 1: whitish-yellow line enters grain; dark blue violet enters oil. Area 2: lemon yellow enters grain; violet blue enters oil. Area 3: orange yellow enters grain; sky blue enters oil. Area 4: reddish orange enters grain; whitish blue enters oil. Area 5: dark reddish brown enters grain; blue white enters oil.

hardly have shifted at all. Microscopic observation of the grain edges while the mount cools to room temperature—that is, while the oil's dispersion curve steadily shifts back from T_5 to $T_4 \cdots$ to T_1—will reveal the gamut of dispersion colors summarized in Fig. 5–9.

The change of refractive index with temperature, dn/dt, varies for different oils. For immersion media of index about 1.633, dn/dt usually approximates — 0.0004; in other words, the oil's refractive index decreases 0.0004 for each 1° C rise in the oil's temperature. For oils in the 1.739 range, dn/dt is usually about — 0.0007. The value dn/dt is generally stated on the label of each oil in the commercially available sets. Precise values for particular oils are listed in the pamphlet available with this text.

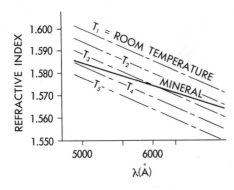

Fig. 5-10. The shift in the dispersion curve of an oil as its temperature is increased from room temperature up to T_5. In contrast, the dispersion curve of a mineral immersed in the oil undergoes no significant shift when subjected to the same temperature increase.

PRACTICAL PROCEDURES

Preparation of Mount

The unknown crystal is first pulverized and the 100 to 200 mesh size is isolated by sieving. (In the case of powdered unknowns, of course, this has already been done). The immersion oil selected should be of refractive index equal to that of the suspected identity of the crystal; if there is no clue as to identity, an oil of index 1.53 should be used. Place a drop of the oil on the slide, dislodging the drop from the glass dropper by tapping it; the dropper should never directly touch the slide, because the immersion media may become contaminated by stray mineral grains that adhere to the dropper on contact. Replace the bottle of immersion oil, *with cap tightly in place*, in the case immediately. Dust a few dozen grains of the unknown crystal onto the oildrop and cover it with a cover glass about $\frac{1}{4}$-by-$\frac{1}{4}$-inch in size.

Microscopic Observations

The grains of the slide mount can now be examined between crossed nicols, using the low-power objective. If the grains are truly isotropic, they will all remain black—that is, extinct—even if observed while the stage is being rotated a full 360 degrees (see pp. 68-70). If the grains brighten, they are anisotropic and present problems of index measurement that will be discussed later. A slightly more sensitive test for anisotropism is to insert the first-order red accessory plate while conducting the preceding observation. The grain is anisotropic if, during rotation, its color varies from the violet background color produced by the plate for the surrounding oil.

After the grain has been verified to be isotropic, oblique illumination may be used to determine whether the grain's index is greater or less than the oil's. An exact match probably will not be obtained the first time, making it necessary to select a second immersion oil and prepare a second mount (on the same glass slide). It is wise to sketch the slide in your notebook (Fig. 5–11) and label each mount with the indices of the oil used. Observations can later be rechecked, for at room temperatures the oils of the mounts (*if under cover glasses*) do not change significantly in index, even after a time lapse of several hours.

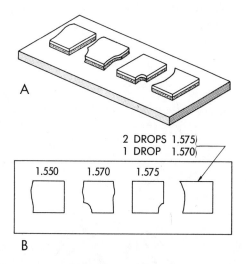

Fig. 5-11. (A) Glass slide with cover glasses over oil mounts. (B) Plane view of slide as sketched in notebook. For each mount the refractive index of the oil used (as given on its label) is recorded. Note that the final matching oil is a mixture of two media of the set.

After the grains have been mounted in a closely matching oil, the relative intensities and dispersion colors of the Becke lines may be utilized to determine the refractive index of the grain for sodium light to within $\pm$ .002, provided the index of the matching oil is corrected for temperature as described in the following section.

Temperature Correction of Index

The label on a bottle of immersion oil usually contains the following information: n_D, its refractive index for sodium light; T, the temperature at which the oil possesses this index; and dn/dt, the amount the index changes per degree change in temperature (see pp. 59-60). The date when the index was measured is, although not often stated, a useful addition to the data on the label. Normally, unless the light source is allowed to heat up the slide mount over an extended period of time, the temperature of the oil in the powder mount will equal room temperature, T_R. If T_R equals T, the index of the oil in the powder mount will be the same as that (n_D) stated on its label. If the two are not equal, then the index of the oil in the powder mount, n_{DM}, must be calculated:

$$n_{DM} = n_D + (T_R - T)\frac{dn}{dt} \qquad \text{(Eq. 5–2)}$$

As an example of the use of Eq. 5–2, assume a mineral has been matched in index to a particular oil ($n_D = 1.530$, $T = 20°$ C, $dn/dt = -.0003$), the room temperature at the time of the match being 25° C. The actual refractive index of the oil (and therefore of the mineral) at the time of the match is

$$n_{DM} = 1.530 + (25° - 20°) \left(\frac{-.0003}{1°} \right) = 1.5285$$

The index reported for the mineral is then rounded off to 1.528 or 1.529; reporting the grain's index as 1.5285 implies a greater accuracy than is attainable by the immersion method ($\pm\ 0.002$). Wilcox (1959, p. 1286)

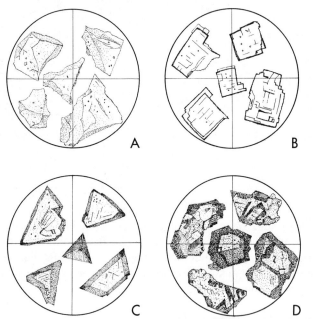

Fig. 5-12. Breakage types commonly seen on crushed isotropic grains: (A) conchoidal fracture but no cleavage; (B) cubic cleavage $\{100\}$—that is, three mutually perpendicular directions of equal ease of cleavage; (C) octahedral cleavage $\{111\}$—that is, four directions of equal ease of cleavage that are parallel to the faces of an octahedron; (D) dodecahedral cleavage $\{110\}$—that is, six directions of equal ease of cleavage parallel to the faces of the dodecahedron. Note that for dodecahedral cleavage the fact of six "competing" directions for cleavage makes it unlikely that a particular direction will be extensively developed; instead, the breakage surface alternately follows one and then another of these six directions. The relief of these grains in oil varies as follows: (A) low, (B) moderate, (C) high, (D) very high.

believes that under favorable circumstances—that is, sodium light, small intervals in refractive index spacings of the media, and close attention to temperature corrections—accuracy of $\pm$ 0.001 may be attainable with Becke line methods. Advanced techniques such as the double variation method of Emmons (1943) permit accuracy to $\pm$ 0.0002.

In addition to measurement of the refractive index, other observations should be made. The color transmitted by the grains should be recorded. Inclusions within the grain—such as empty bubbles, liquid-filled bubbles, tiny crystals, cavities where tiny crystals once were (that is, negative crystals), or crystallites—should be searched for, using high-power magnification. Linear parallelism of elongated inclusions or of bubble trains should be noted. The nature of the cleavage or fracture of the crushed grains may also yield information valuable in identification of the mineral. Fig. 5–12 illustrates the grain shapes typical of the different types of breakage observable for isotropic crystals.

Measurement of Specific Gravity

The specific gravity of minerals can sometimes be determined within fairly narrow limits by noting whether they sink or float in the immersion oil of the mount. To facilitate this observation, Foster (1947, p. 463) suggests using a glass slide upon which two strips of a second slide have been cemented with water glass (Fig. 5–13). A few drops of liquid in the trough

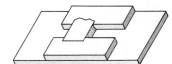

Fig. 5-13.   Glass slide with oil trough. (After Foster, 1947.)

thus formed, a pinch of mineral grains, and emplacement of a cover slip produce a mount of sufficient oil depth to permit easy discrimination between grains resting on the bottom and those floating on top; the high-power objective is best for this examination. Alternatively, coarser grains of ground glass may be added to an ordinary oil mount in addition to the smaller mineral grains. In this case, the coarser grains prop up the cover glass to produce again an increased thickness of oil.

B. M. Shaub (1959, p. 890) suggests that the microscope stage be tilted from its horizontal position by 60 degrees or more. In viewing an ordinary powder mount as the stage, so tilted, is rotated, the smaller particles are observed to rise or sink depending upon their density relative to that of the immersion oil. Since the polarizing microscope inverts the image, grains more dense than the liquid appear to rise in the field of view.

The immersion oils used in the refractive index range 1.633 to 1.739 commonly consist of mixtures of α-monochloronaphthalene (sp gr at

$20° = 1.194$) and diiodomethane (sp gr at $20° = 3.325$). The relation between refractive index and density for a typical set of such mixtures is presented in Fig. 5-14.

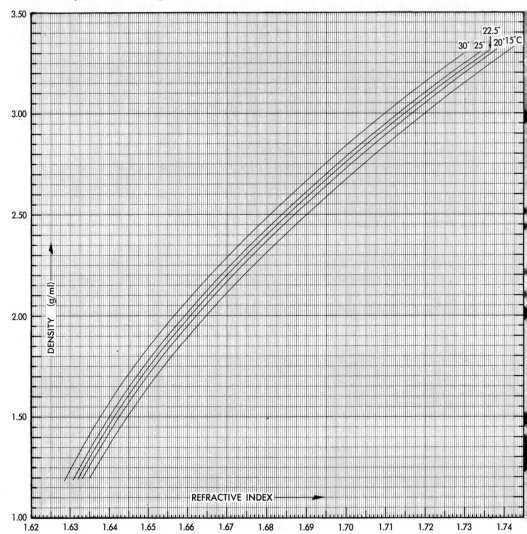

Fig. 5-14. Relation between refractive index and density for mixtures of diiodomethane and α-monochloronaphthalene.

MINERAL IDENTIFICATION

After the refractive index n has been measured for an unknown mineral, consult Table II–1 (Appendix II, p. 238) to determine, if possible, the identity of the mineral.

SIX   Optical Indicatrices And Ellipses

REVIEW OF TERMINOLOGY

A ray of light represents the path by which a continuous, infinitely thin, straight-line stream of energy travels outward from a source. The energy travels along this ray as a series of vibrations that in isotropic media are perpendicular to the ray path (Fig. 6–1A) but in anisotropic media generally are not (Fig. 6–1B). A wave normal direction (W. N. in Fig. 6–1) may be defined as one that (1) lies in the same plane as the ray path and vibration direction and (2) is perpendicular to the vibration direction. Only for light traveling in isotropic media do the wave normal and ray directions invariably coincide; within anisotropic media they generally do not.

A   B

Fig. 6-1. Geometric relationships between ray path (heavy arrow), wave normal (dashed line), and vibration directions (small arrows). (A) For light in isotropic media; (B) for light in anisotropic media. All lines are in the plane of the paper. The wave normal (W.N.) is perpendicular to the vibration directions.

To speak in terms of one ray is, of course, unrealistic. The thinnest pencil of light one can isolate will not be infinitely thin; rather it will consist of numerous rays. The points on these rays that, at a precise moment of time, are in the same stages of vibration (for example, points *a, b, c, d,*

65

and *e* of Fig. 6–2A) are said to be in phase with each other. A surface connecting such points for numerous rays constitutes a wave surface or wave front. The vibration directions of the rays are always parallel (or

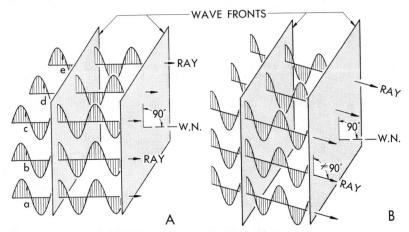

Fig. 6-2. Geometric relationships between wave fronts, wave normals, and ray paths in (A) isotropic media and (B) anisotropic media. For simplicity only polarized light is considered. In (B) the ray paths are not perpendicular to the wave fronts.

tangent) to this wave front; the wave normal is perpendicular to it. Ray paths are generally not perpendicular to the wave front except in isotropic media and special cases in anisotropic media wherein they happen to coincide with the wave normal.

GENERAL CONCEPT OF THE INDICATRIX

The optical indicatrix illustrates how the refractive index of a transparent material varies according to the vibration direction of the light wave in the material (monochromatic light assumed). Consider an infinite number of vectors radiating outward in all directions from a common point within the crystal. Each vector is drawn proportional in length to the crystal's refractive index for light vibrating parallel to that vector direction. The indicatrix is a surface connecting the tips of these vectors. The vectors themselves are generally omitted when this connecting surface is drawn, but in Fig. 6–3 a few are shown for illustration. The indicatrix is purely a method of rationalizing optical phenomena. As such, it furnishes an orderly framework whereby the optical phenomena associated with transparent crystals may be interpreted, remembered, and predicted. It is particularly useful for anisotropic crystals. For isotropic media, however, indicatrix theory is not particularly advantageous, chiefly because of the simplicity of their optical behavior. We discuss the theory here because it serves as an introductory step in the development of the highly important anisotropic indicatrices.

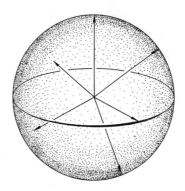

Fig. 6-3. Isotropic indicatrix of a crystal for sodium light. The crystal's refractive index for sodium light (n_D) remains constant regardless of the direction in which the light is vibrating; the indicatrix is therefore a sphere.

THE ISOTROPIC INDICATRIX

Description

In isotropic media, by definition, the index of refraction does not change with the vibration direction of the light. Consequently, all the vectors relating refractive index to vibration direction are of equal length, and therefore all isotropic indicatrices are perfect spheres (Fig. 6–3). Transparent glasses, liquids, and isometric crystals—not under strain—are characterized by such indicatrices.

As discussed under the topic of dispersion, the index of an isotropic substance varies according to the wavelength of light used. Sodium chloride, for example, possesses the indices $n_C = 1.541$, $n_D = 1.544$, and $n_F = 1.553$. Technically, therefore, a slightly different-sized indicatrix exists for each wavelength of light. Fig. 6–4 illustrates sodium chloride's indicatrices for C-, D-, and F-light, respectively. For intermediate wavelengths the indicatrices would be intermediate to those shown.

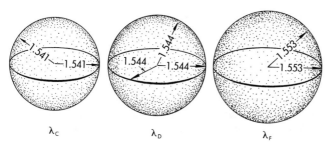

Fig. 6-4. Comparison of the optical indicatrices of crystalline sodium chloride for C, D, and F wavelengths. Dimensional changes between the three spheres are exaggerated for illustrative purposes.

Application

The incidence of light upon the surface of an isotropic material makes few demands on the indicatrix theory. The crystal surface in question is assumed to pass through the indicatrix center, and the resultant intersection between surface and indicatrix is a circle of radius proportional to the crystal's refractive index for the light. In indicatrix theory such a circular

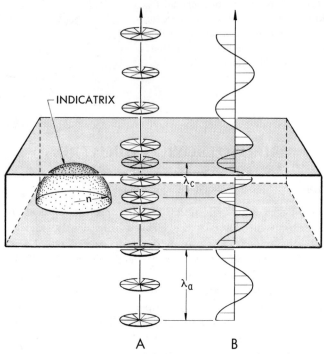

Fig. 6-5. The effect of an isotropic plate upon (A) a normally incident unpolarized ray and (B) a polarized ray. The lower crystal surface intersects the spherical isotropic indicatrix in a circle whose radius, n, equals that of the sphere, n being thus the refractive index of the plate. Changes in amplitude due to reflectional losses at interfaces, etc., are not considered in the drawing. Note that the wavelength of the light while in the crystal (λ_c) is smaller than its wavelength in air (λ_a).

section indicates that the crystal "permits" the entering light to vibrate within the crystal in the same direction(s) as it did prior to entry; that is, the light is not required to vibrate parallel to a particular direction in order to pass through the crystal. Thus, unpolarized light (Fig. 6–5A) remains unpolarized after entry into the crystal whereas polarized light (Fig. 6–5B) maintains its same plane of polarization after entry.

Since isotropic materials cannot alter the direction of polarization of the entering light, an isotropic grain, viewed between crossed nicols, always

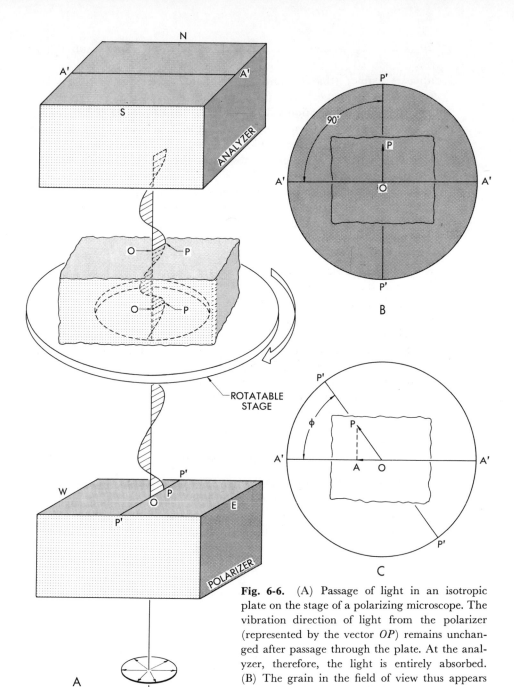

Fig. 6-6. (A) Passage of light in an isotropic plate on the stage of a polarizing microscope. The vibration direction of light from the polarizer (represented by the vector OP) remains unchanged after passage through the plate. At the analyzer, therefore, the light is entirely absorbed. (B) The grain in the field of view thus appears black (extinct) when viewed between crossed nicols. Vectorially OP represents the amplitude and direction of vibration of light from the polarizer both before and after passage through the crystal. OP has no component parallel to $A'A'$, the privileged direction of the analyzer, if the nicols are crossed. Thus no light is transmitted through the analyzer. (C) If either the analyzer or polarizer is rotated in its holder so as to cause ϕ, the angle between their privileged directions, to be other than 90°, then there is a component of OP (that is, OA) that is parallel to $A'A'$, the privileged direction of the analyzer. Consequently, the analyzer transmits a component of the light, and the grain no longer appears extinct.

appears extinct (that is transmits no light), even during a 360-degree rotation of the stage.* This property thus serves as a test for isotropism. The following explanation will clarify: The light from the polarizer passes through the crystal with its N–S vibration direction unchanged (Fig. 6–6A). At the analyzer, however, which is usually set to transmit only E–W vibrating light, all this light is absorbed, the analyzer acting as though opaque to this light. The situation is considered vectorially in Fig. 6–6B. Here OP represents the amplitude and direction of the light vibrations from the polarizer that have passed through the crystal. However, vector OP has no component parallel to $A'A'$, the direction of light vibration transmitted by the analyzer. Consequently, not even the smallest component of the light energy vibrating parallel to OP is transmitted by the analyzer. Rotation of the stage changes neither the polarizer's privileged direction $P'P'$ nor the analyzer's privileged direction $A'A'$. Thus the vectorial relationships shown in Fig. 6–6B remain the same, regardless of the position of the stage.

On the other hand, if either the polarizer or analyzer is rotated so as to change ϕ (phi), the angle between $P'P'$ and $A'A'$ (Fig. 6–6C), to a value other than 90 degrees, the isotropic grain will no longer appear extinct between the two nicols. Instead, a vector component of OP (OA in Fig. 6–6C) will be transmitted by the analyzer. Note that the amplitude of OA may be graphically determined from that of OP by dropping a perpendicular from point P to line $A'A'$; expressed mathematically the relationship is

$$OA = \cos \phi \cdot OP \qquad \text{(Eq. 6–1)}$$

The greater the amplitude of vibration of the light, the brighter the light. In the case of Fig. 6–6C, the light passing through the crystal (amplitude OP) is brighter than that passing through the analyzer (amplitude OA). Only when the angle ϕ of Eq. 6–1 equals 0 degrees will OP and OA be equal and therefore represent equal brightnesses.

THE UNIAXIAL INDICATRIX

Historical Origin

Description and discussion of all the steps that prefaced the elucidation of the indicatrix theory by Fletcher (1891) cannot be included here. We will, however, present in the following sections a few of the major discoveries that indicated the existence of other than isotropic materials (and, subsequently, the need of more than Snell's law to explain their optical behavior).

Discovery of O and E Rays. Erasmus Bartholinus in 1669 reported that a dot on paper (or a ray of light) gave rise to two images when viewed

---

* The possibility of optical activity is here disregarded.

through a rhomb of calcite. Thus a normally incident parent ray (for example, PO in Fig. 6–7) produced not only ray OP_O but also ray OP_E, the two rays following different paths within the crystal. Ray OP_O was called the *ordinary ray* or *O ray* (since its path was predictable by Snell's law) whereas ray OP_E was called the *extraordinary ray* or *E ray* (because its path was an obvious contradiction of Snell's law). This contradictory nature of path OP_E was obvious since $\angle r$ did not equal 0 degrees as required by Snell's law.

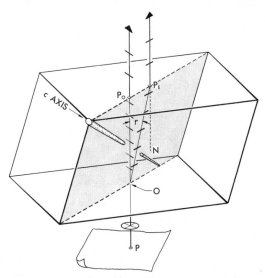

Fig. 6-7. Double refraction of an unpolarized ray, PO, by a rhomb of Iceland spar (clear calcite) to produce two rays, OP_O and OP_E. If the rhomb is viewed from above, two images of point P are seen, one from rays emergent at P_O, the second from rays emergent at P_E. The angle between rays OP_O and OP_E is exaggerated for illustrative purposes. The construction line, NP_E, is perpendicular to the upper face of the rhomb.

Fresnel and Arago (1811) showed that the O and E rays in calcite were polarized at right angles to each other. The light composing the O ray was always observed to vibrate at right angles to the plane containing both the O-ray path and the c axis (that is, it vibrated perpendicular to the shaded plane in Fig. 6–7). The light composing the E ray, on the other hand, was always observed to vibrate *within* this plane. Moreover, whenever a parent ray is normally incident on the surface of an anisotropic crystal, the vibrations of the refracted rays entering this crystal are always parallel to this surface. Thus, in Fig. 6–7, the E and O vibrations are both parallel to the underside of the rhomb.

Limitations of Snell's Law. Although Snell's law cannot correctly predict the E-ray path, it does determine its associated wave normal direction. Fig. 6–8 is a cross-sectional view of an unpolarized ray PO passing from an isotropic medium (refractive index: n_i) into an anisotropic medium thus to form two refracted rays, ordinary ray OP_O and extraordinary ray OP_E. The vibration directions of the light for both rays are shown, those

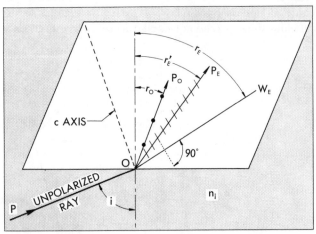

Fig. 6-8. Cross-sectional view of the passage of a ray of unpolarized light, PO, from an isotropic medium (shaded) into a uniaxial crystal to form an ordinary ray, OP_O, and an extraordinary ray, OP_E. The E-wave normal OW_E is a direction perpendicular to the vibration directions drawn along OP_E; to emphasize this, one of these vibrations is extended by dotted lines.

associated with OP_E as short lines and those for OP_O as dots (since the latter represent vibrations precisely perpendicular to the cross-sectional plane). As will be later discussed more fully, the crystal exhibits different indices of refraction for these two different vibration directions, an index of ω for the O-ray vibrations and of ϵ' for the E-ray vibrations. Snell's law holds for the ordinary ray, thus

$$n_i \sin i = \omega \sin r_O$$

but it does not for the extraordinary ray since

$$n_i \sin i \neq \epsilon' \sin r'_E$$

However, if the angle of refraction is measured as r_E—that is, with respect to direction OW_E in Fig. 6–8 rather than to the E-ray path—Snell's law holds:

$$n_i \sin i = \epsilon' \sin r_E \qquad \text{(Eq. 6–2)}$$

OW_E is, of course, the direction perpendicular to those vibration directions marked off along OP_E; thus it is the E-wave normal.

The limitations in the applicability of Snell's law may now be summarized. Snell's law predicts ray paths correctly only if such paths happen to coincide with their wave normals; in the prediction of wave normal directions the law is infallible. Such coincidence of ray path and wave normal occurs if the light composing the ray vibrates perpendicular to the ray's path; this situation occurs for (1) all rays in isotropic media and (2) the ordinary ray in uniaxial media but generally not for (3) the E ray in uniaxial media. Thus the E-ray path in uniaxial media does not follow Snell's law; however, its associated wave normal does.

Birefringence. If a calcite rhomb $\frac{3}{4}$-inch thick or more is placed over a dot on a sheet of paper (Fig. 6–9), the dot image P_O' formed by the O rays

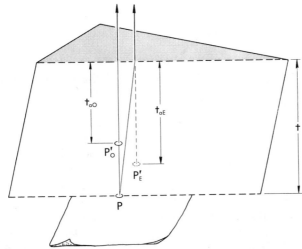

Fig. 6-9. The front half of the rhomb of Fig. 6-7 has been removed to expose the shaded plane shown therein. P_O' and P_E' represent the apparent positions of the O- and E-ray images of dot P as seen by an eye looking through the rhomb. The apparent thickness of the rhomb for the O- and E-ray images of P (that is, t_{aO} and t_{aE} respectively) are compared with the rhomb's true thickness, t.

appears to be shallower than P_E', the dot image formed by the E rays. Thus the rhomb's apparent thickness for the O ray (that is, t_{aO}) is less than that for the E ray (that is, t_{aE}). Symbolizing the crystal's refractive index for the O ray as ω and that for the E ray as ϵ', it is apparent from the Duc de Chaulnes equation (Eq. 5–1) that $\omega = t/t_{aO}$ and $\epsilon' = t/t_{aE}$, where t signifies the true thickness. Consequently, the calcite rhomb has two indices of refraction, the larger being associated with the O ray. Possession of more than one index of refraction is known as birefringence; only birefringent crystal plates are capable of producing double refraction— that is, doubling of images.

If a clear calcite rhomb is available, the reader is invited to verify the above observations and conclusions as an exercise. As a first step, the O dot may be recognized since it remains stationary even if viewed as the crystal is rotated on an axis parallel to PP'_O (Fig. 6–9). The relative depths of the E- and O-dot images may be qualitatively determined by viewing them while moving the head from side to side, cobralike. Because of the effect of parallax, the shallower O-dot image will appear to move from side to side with respect to the E-dot image, the direction of its apparent motion being opposite to that of one's head. (The principle of parallax can be quickly demonstrated by lining up two vertically held pencils, one 10 inches from your eyes, the other at arm's length. If you now move your head to either side of this alignment, the closer pencil appears to move with respect to the more distant one, its motion being opposite to that of the head.)

Description and Discussion

In anisotropic media the index of refraction actually varies according to the vibration direction of the light in the crystal. Consequently, the optical indicatrix for anisotropic media is not a sphere but an ellipsoid. Of the two types of anisotropic optical indicatrices, the uniaxial indicatrix will be discussed first, since it is simpler.

Crystals of the hexagonal and tetragonal systems exhibit, for mono-chromatic light vibrating parallel to the c axis, a unique index of refraction customarily symbolized as ϵ. In Fig. 6–10A, therefore, a vector proportional in length to the value of ϵ has been drawn parallel to the c axis to indicate this. On the other hand, for all vibration directions at 90 degrees to the c axis, the crystal's refractive indices all equal a common value symbolized as ω.* By constructing vectors proportional in length to ω along these vibration directions, a circle of radius ω is defined (Fig. 6–10A); this circular section is, of course, always perpendicular to the c axis.

For light vibrating at a random angle θ to the c axis (Fig. 6–10A), the crystal exhibits a refractive index somewhere between ω and ϵ in value; these intermediate indices are symbolized as ϵ'. The value of ϵ' can be computed from the formula

$$\epsilon' = \frac{\omega \epsilon}{\sqrt{\omega^2 \cos^2 \theta + \epsilon^2 \sin^2 \theta}} \qquad \text{(Eq. 6–3)}$$

For calculation purposes this equation may be rewritten

$$\epsilon' = \frac{\omega}{\sqrt{1 + \left(\frac{\omega^2}{\epsilon^2} - 1\right) \cos^2 \theta}} \qquad \text{(Eq. 6–4)}$$

Plane 1 in Fig. 6-10B indicates by the vector lengths how, as θ (for a vibration direction) varies from θ_1 to θ_3, the corresponding crystal index varies

---

* Other symbols sometimes used for ϵ and ω are

$\epsilon: n_\epsilon; \mathcal{N}_\epsilon; E; n_E; \mathcal{N}_E$

$\omega: n_\omega; \mathcal{N}_\omega; O; n_O; \mathcal{N}_O$

from ϵ_1' to ϵ_3'. The tips of all such vectors within plane 1 fall along an ellipse (for which Eq. 6–3 is the equation in polar coordinates). Similarly, within any other plane containing the crystal's c axis (for example, plane 2) the vectors

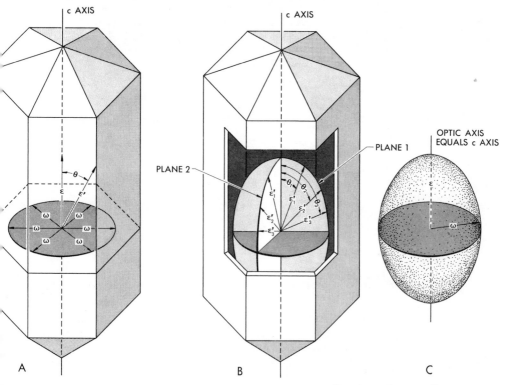

Fig. 6-10. (A) Angular relationships to the c axis of the vibrations corresponding to indices ϵ, ω, and of one particular value of ϵ'. All ω vibrations lie in the plane perpendicular to the c axis. (B) Vector lengths in planes 1 and 2 indicate the variation in index ϵ' of the crystal for light vibrating parallel to them. Their tips outline identical ellipses in all planes through the c axis (for example, planes 1 and 2). (C) The uniaxial indicatrix for the crystal.

describing the ϵ' values fall along an ellipse identical with that for plane 1. Thus the ellipse of plane 1 can be rotated about the c axis as a hinge to coincide with any of these identical ellipses. As a result an ellipsoid of rotation (Fig. 6–10C) is formed. The length of any radius of this ellipsoid indicates the crystal's refractive index for light vibrating parallel thereto. Fig. 6–10C is called the indicatrix or, more precisely, the uniaxial indicatrix (since there is only one axis in it that is perpendicular to a circular section).

The indices ω and ϵ represent the maximum and minimum—or minimum and maximum—refractive indices measurable in a tetragonal or hexagonal crystal. Collectively they may be referred to as the crystal's principal indices. Uniaxial materials fall naturally into two categories: uniaxial positive, in which by definition the value $(\epsilon - \omega)$ is positive in sign, and uniaxial negative crystals, in which $(\epsilon - \omega)$ is negative in sign (that is, $\epsilon < \omega$). If the indicatrix for each of these two types is drawn in its customary orientation—that is, with the c axis vertical (Fig. 6–11)—the

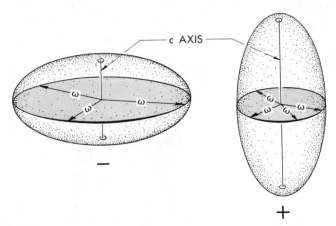

Fig. 6-11. Comparison of the positive and negative uniaxial indicatrices. The circular section for each is shaded dark gray.

positive indicatrix (a prolate spheroid) may be mnemonically linked to the vertical stroke of a $+$ sign and the negative indicatrix (an oblate spheroid) to a $-$ sign. Note that regardless of sign, there is only one circular section in a uniaxial indicatrix; its radius is always equal to principal index ω; and it lies in a plane normal to the c axis.

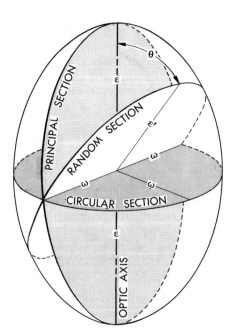

Fig. 6-12. Illustration of the three types of central sections through a uniaxial indicatrix.

Directions and Central Sections

Terminology. The direction in the uniaxial indicatrix that coincides with the crystal's c axis is called the *optic axis* (heavy vertical line in Fig. 6–12). With reference to it, three types of sections may be cut through the indicatrix center: (1) *a principal plane, principal section, or principal ellipse* (that is, one which contains the optic axis and therefore intersects the indicatrix as an ellipse whose semiaxes* are equal to ϵ and ω); (2) the previously mentioned *circular section* (that is, a section cut normal to the optic axis); and (3) a *random section* (that is, the intersection of the indicatrix with a plane cut at a random angle (θ) to the optic axis). A random section always intersects the indicatrix in an ellipse whose semiaxes are ω and ϵ'; the precise value of ϵ' can be calculated from Eq. 6–4, if the value of the angle (θ) is known.

Wave Normals, Ray Paths, and Vibration Directions

For a random wave normal direction in a uniaxial indicatrix (for example OW in Fig. 6–13A), the two associated vibration directions lie in the plane perpendicular to this wave normal. This plane, extended outward from the indicatrix center, intersects the indicatrix in an ellipse (stippled) whose major and minor axes (OV_E and OV_O) constitute the only two vibration directions associable with wave normal OW. Consequently, light with a random wave normal direction such as OW would be constrained to vibrate parallel to OV_O or OV_E or both while passing through a uniaxial crystal. One of these two vibration directions, OV_O, will always be perpendicular to the optic axis, being the intersection of the stippled plane normal to OW with the circular section (shaded dark gray in Fig. 6–13A) and thus a radius of the circular section. The second of these vibration directions, OV_E, always lies within the same plane (shaded gray in Fig. 6–13B) as the optic axis and the wave normal OW.

Only one ray path can be associated with a given combination of wave normal and vibration direction. All three (path, normal, and vibration direction) lie within a common plane that usually intersects the indicatrix in an ellipse. Within this ellipse the ray path and vibration direction represent conjugate radii—an important point to understand. To illustrate, the ray path associated with wave normal OW and vibration direction OV_E also lies in the ellipse shaded gray in Fig. 6–13B, being OR_E, the radius conjugate to OV_E in this ellipse. Similarly, the ray path associated with wave normal OW and vibration direction OV_O, lies in the unshaded ellipse in Fig. 6–13B; since OV_O is a semiaxis of this ellipse, OR_O, the ray path and

---

* The term semiaxes is used here to refer collectively to the semimajor and semiminor axes of an ellipse. For convenience, a brief summary of the nomenclature, properties, and constructions associated with ellipses, pertinent to optical crystallography, is given in Appendix I. This section, particularly where it deals with the concept of conjugate radii of an ellipse, should be well understood before reading the following paragraphs.

radius conjugate to it, is perpendicular to OV_O.* Hence OR_O coincides with wave normal OW, which in turn is always perpendicular to vibration direction OV_O. Thus, ray OR_O, since it vibrates perpendicular to and coincides in path with its wave normal direction, is an ordinary ray and obeys Snell's law. Ray OR_E, however, does not conform to this and is thus an extraordinary ray whose path does not obey Snell's law. The significance of the subscripts E and O used in the preceding discussions (and that to follow) becomes apparent. Subscript E refers to the vibration direction and ray path of the extraordinary ray; subscript O to those for the ordinary ray.

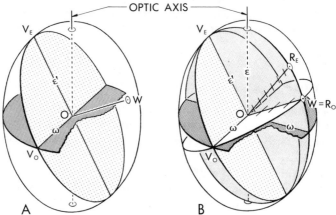

Fig. 6-13A-B.   (A) Relation of a random wave normal OW to the only two vibration directions, OV_O and OV_E, that are associable with it. The stippled plane perpendicular to OW intersects the indicatrix in an ellipse for which OV_O and OV_E are semiaxes. (B) Location of OR_E, the only ray path associable with OW and OV_E. All three lie in the same plane (shaded gray). Similarly, ray path OR_O, the only path associable with OW and OV_O, lies in the same plane (unshaded) with them; OR_O coincides with OW in this instance. In both (A) and (B) the circular section is shaded dark gray.

The ray path and wave normal direction for light in a uniaxial crystal coincide only if their associated vibration direction is perpendicular or parallel to the optic axis. A vibration direction thus oriented will correspond to a principal index, ω or ϵ. In consequence, this vibration direction will be the longest or shortest radius (that is, a semiaxis) for all elliptical cross sections of the indicatrix in which it may lie. The case for a vibration direction that is perpendicular to the optic axis has already been illustrated, vibration direction OV_O in Fig. 6–13B having been associated with a ray path (OR_O) and wave normal (OW) that coincided in direction.

* The radius conjugate to one semiaxis of an ellipse is the other semiaxis, the semiaxes thus constituting the only pair of mutually perpendicular conjugate radii in an ellipse. The reader should confirm this.

The directional relationships of ray paths and vibration directions to wave normals that are at the special angles of 90 or 0 degrees to the optic axis may be determined in a manner similar to that for a random wave normal. For example, if wave normal OW is at 90 degrees to the optic axis (Fig. 6–13C), the two vibration directions are OV_O and OV_E (for which the indices are ω and ϵ, respectively). Unlike the random case, however, the E-ray path now also coincides with its associated wave normal OW (since its vibration direction OV_E and wave normal OW are conjugate radii within the gray shaded ellipse). On the other hand, if the given wave normal

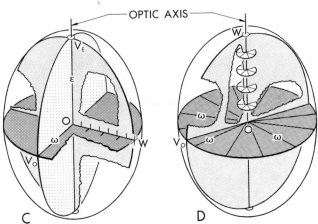

Fig. 6-13C-D. (C) Relation of vibration directions (OV_O and OV_E) and ray paths (these coincide with OW) to OW, a wave normal perpendicular to the optic axis. Note coincidence of O- and E-ray paths. (D) Relation of vibration directions (which are all possible radii of the heavily shaded circular section) and ray path (coincident with OW) to a wave normal, OW, parallel to the optic axis. In both (C) and (D) the circular section is shaded dark gray.

coincides with the optic axis (Fig. 6–13D), the indicatrix section perpendicular to OW is a circular section. Consequently, no particular privileged directions are associable with this wave normal; instead, a wave traveling along the optic axis may vibrate parallel to any one of the innumerable radii of the circular section—or to all of them if the incident ray was unpolarized. Only the index ω can be associated with this ray, and its path coincides with the wave normal OW. The coincidence of ray path and wave normal for each vibration in Fig. 6–13D is readily understood if a principal ellipse (shaded gray) is drawn through one of them (for example, OV_O). It is now obvious that OV_O and OW are semiaxes and therefore conjugate radii of this ellipse; consequently, OW represents a ray path as well as a wave normal direction.

In summary, it may be noted from Fig. 6–13B that, with respect to a wave normal at a random angle to the optic axis, either of two mutually perpendicular vibrations and, therefore, either of two ray paths may be associated. Both possible ray paths lie within the principal plane containing the wave normal (shaded gray in Fig. 6–13B). One of these rays, the O ray, vibrates normal to this principal plane; the second, the E ray, vibrates within this principal plane along the radius of the ellipse that is conjugate to its path.

Application to Light Incidence on Crystal Surfaces

Types of Incidence. A light ray entering a uniaxial crystal defines a plane of incidence (see p. 8) that is either parallel or at an angle to the crystal's optic axis. Rays 1 and 2 in Fig. 6–14 illustrate the first case; both

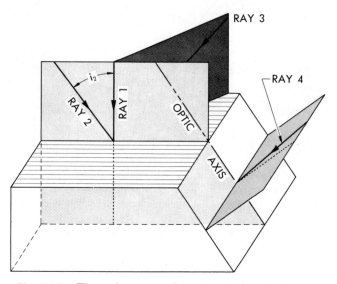

Fig. 6-14. The various types of angular relationships between the optic axis of a uniaxial crystal and the planes of incidence defined by light rays entering it. The lightly shaded plane of incidence containing rays 1 and 2 is parallel to the optic axis. The heavily shaded plane of incidence containing ray 3 represents the general case, being at an angle other than 0 or 90 degrees to the optic axis. The moderately shaded plane of incidence containing ray 4 is at 90 degrees to the optic axis; this 90-degree angle can occur only for incidence upon a crystal surface parallel to the optic axis. The normals to the crystal faces are shown by dotted lines.

lie in a plane of incidence that parallels the optic axis. Ray 1 is further specialized since it coincides with the normal to the crystal face; it therefore represents what is called the case of *normal incidence* upon this crystal face. Ray 2, on the other hand, is at an angle of i_2 to the normal. Ray 3 illustrates the more general case since it defines a plane of incidence (heavily shaded) that is at an angle to the crystal's optic axis. Only if the optic axis is parallel to the crystal face upon which incidence occurs can the angle between the plane of incidence and the optic axis ever equal 90 degrees (cf. the plane of incidence containing Ray 4 in Fig. 6–14).

Rays 1, 2, 3, and 4, after entering the crystal, will generally form an O and E ray that travel along different paths within the crystal. For all four cases the O-ray path may be readily determined since (1) it always lies within the plane of incidence and (2) it follows Snell's law. The path of the E ray is rather difficult to determine for ray 3, moderately so for ray 2, and simpler for ray 4 and for normal incidence such as ray 1 represents. In this text, the method of locating the E-ray path will be discussed only for the cases represented by ray 1 (the case of normal incidence), by ray 2, and by ray 4. The case of normal incidence in particular is most important to understand in the routine practice of optical crystallography and will therefore be next discussed in greatest detail.

Normal Incidence. Prescribed geometrical relationships exist between the ray path, wave normal, and vibration direction of a given light energy traveling within a uniaxial crystal (p. 77). Thus a refracted ray path in a uniaxial crystal can be determined if its associated wave normal and vibration direction can be located. For normal incidence ($i = 0$ degrees), the wave normals of the rays entering the crystal are readily located, for they are perpendicular to the crystal surface upon which the incidence occurs (since for the revised Snell's law, if $i = 0°$ for the incident wave normal, then $r = 0°$ for all refracted wave normals). Therefore, because vibrations are invariably perpendicular to wave normals, *the vibration directions for the rays entering the crystal are always parallel to the crystal surface upon which normal incidence occurs.* The precise directions of these vibrations can be visualized if the crystal's indicatrix can be imagined with its center on the crystal face. As illustrated in Fig. 6–15, the intersection between the indicatrix and crystal face is either an ellipse (cf. faces m and q) or a circle (cf. face c), depending upon the angle between the face and the optic axis. If the intersection is an ellipse, its semiaxes mark the vibration directions, after entry into the crystal, of light normally incident upon the face. If the intersection is a circle, no privileged directions exist; thus, as for isotropic media, light normally incident on this face will vibrate in the same direction(s) after entry into the crystal as it did prior to entry.

Fig. 6–15 further illustrates that it is permissible to draw the indicatrix *anywhere* within (or outside) the crystal just so long as its optic axis is maintained parallel to the crystal's c axis. Thus the indicatrix has a specific orientation but not a specific location within the crystal.

Section perpendicular to the optic axis. A section cut perpendicular to the optic axis of a uniaxial crystal (plane c in Fig. 6–15) intersects the indicatrix in a circle of radius ω. The situation is thus analogous to that described for the isotropic plate (p. 68); that is, the entering light may be considered to vibrate with equal ease, after entry into the crystal, parallel to any (or all) radii of the circle of intersection. Consequently, unpolarized

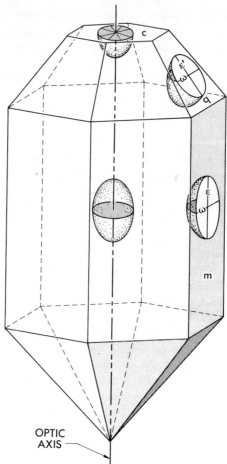

Fig. 6-15. Types of intersection between the faces of a crystal and its indicatrix. The heavy radii of the ellipse (or circle) indicate the crystal's privileged directions for light entering by normal incidence on this face.

light normally incident on the plate will remain unpolarized (for example ray *BB'* in Fig. 6–16) whereas a plane-polarized ray (for example, *CC'* in the figure) will retain its same direction of polarization.* Thus, like the isotropic plate (and for the same reasons), this plate viewed between crossed nicols would remain completely extinct, even if observed during rotation of the stage, particularly if the substage iris diaphragm is closed down to a small aperture.

The analogy between this particular section and an isotropic plate may be carried even further since it too lacks birefringence, exhibiting only one index, ω, for normally incident light. (All other types of sections through

---

* The possibility of rotatory polarization is disregarded here.

the indicatrix, we shall soon see, do exhibit birefringence.) Furthermore, the rays BB' and CC' act as ordinary rays since they coincide with their wave normal directions (or, stated alternatively, since their associated

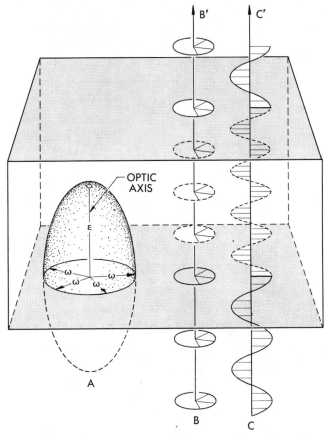

Fig. 6-16. Normal incidence of light upon the under surface of a crystal that was cut perpendicular to its optic axis. The orientation of the crystal's optical indicatrix to this surface is shown at A. The passage through the crystal of an unpolarized ray, BB', and of a polarized ray, CC', is illustrated.

vibration directions are perpendicular to their path). Thus these rays follow Snell's law, even in its unrevised form. Such isotropic-like phenomena are observed in a uniaxial crystal whenever a ray travels along the optic axis (that is, the c axis) of the crystal. For this reason the optic axis is sometimes called the *axis of isotropy*.

Section parallel to the optic axis. In this case (Fig. 6–17) the uniaxial indicatrix, if centered on the crystal's boundary plane, intersects this plane in an ellipse whose semimajor and semiminor axes equal ϵ and ω, the two extremes of refractive index (that is, principal indices) that a uniaxial crystal can exhibit. These two axes also represent the two mutually perpendicular directions (that is, privileged directions) parallel to which the light is constrained to vibrate while passing through the crystal. Consequently, an unpolarized light ray normally incident upon the plate (A in Fig. 6–17) will, after entry, be resolved into two different rays. They travel, in this case, along a common path but vibrate perpendicular to each other. For the ray vibrating parallel to the ω privileged direction, the crystal exhibits the index ω, for the ray vibrating parallel to the ϵ

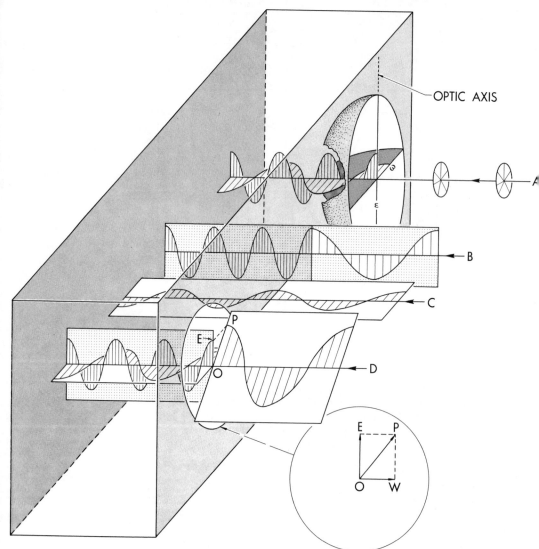

Fig. 6-17. Normal incidence of an unpolarized ray, A, and of polarized rays, B, C, and D, on a crystal face parallel to the optic axis. Circular inset: enlarged, face-on view of the resolution of vector OP upon entering the crystal.

privileged direction, the crystal exhibits the index ϵ. Thus, this plate is birefringent, exhibiting two different refractive indices, ϵ and ω.

Plane-polarized light normally incident on this same crystal plate (Fig. 6–17) will be differently affected according to the angle between its plane of polarization and the privileged directions of the plate. If the incident light is plane polarized parallel to one of the privileged directions, the crystal transmits only one ray; this single ray vibrates parallel to this particular privileged direction only. Examples are rays B and C in Fig. 6–17, ray B vibrating parallel to the ϵ privileged direction, ray C to the ω privileged direction; in each case the crystal exhibits only one refractive index (ϵ for ray B, ω for ray C).

If, however, the incident light is plane polarized parallel to neither privileged direction (ray D in Fig. 6–17), it is resolved vectorially into two rays upon entering the crystal, each ray vibrating parallel to a privileged direction. Fig. 6–18 illustrates in greater detail the resolution of the incident

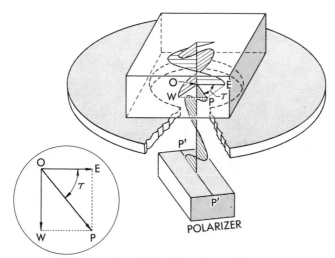

Fig. 6-18. Resolution of light from the polarizer upon entry into the crystal. Circular inset: vertical and enlarged view of the resolution of OP into two components that occurs as the light enters the under surface of the crystal.

light's amplitude (OP) into the two amplitudes (OE and OW) associable with these two rays. The relative magnitudes of these amplitudes are

$$OE = OP \cos \tau$$

$$OW = OP \sin \tau$$

where τ (tau) is defined as the angle between OP and the privileged direction OE for the extraordinary ray in the crystal.

The light intensity of a ray is proportional to the square of its amplitude. Thus in Fig. 6–18

$$I \quad \text{is proportional to} \quad OP^2$$
$$I_E \quad \text{is proportional to} \quad OP^2 \cos^2 \tau$$
$$I_O \quad \text{is proportional to} \quad OP^2 \sin^2 \tau$$

where I, I_E, and I_O respectively represent the intensities of the incident, extraordinary, and ordinary rays. As may be seen in the figure, the rotating stage of a polarizing microscope permits τ to be varied between 0 and 90 degrees. Consequently, all the incident light energy may alternatively be concentrated in the extraordinary ray or in the ordinary ray, or be divided equally between them ($\tau = 45°$) or in any proportion. [What is the intensity ratio of the extraordinary ray to the ordinary ray if $\tau = 30°$?]

Section cut at a random angle (θ) *to the optic axis.* For this orientation, if the crystal's indicatrix is centered on the boundary plane (darkest shading in Fig. 6–19), their intersection is an ellipse (vertically ruled)

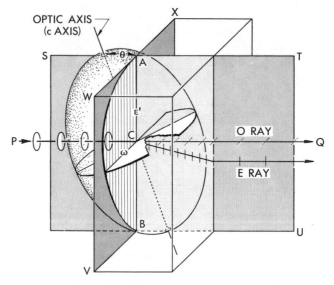

Fig. 6-19. Normal incidence of a ray of unpolarized light, *PC*, on a random section, *XWV*, of a uniaxial crystal. Note its bifurcation within the crystal into an *E* and an *O* ray, each lying within plane *STU*. Ray *PC* and the crystal's optic axis also lie in plane *STU*.

whose major and minor axes mark two privileged directions corresponding to indices ω and ϵ' in the crystal plate. As shown, the normally incident, unpolarized, light ray *PC* is separated into an *O* and an *E* ray vibrating parallel to the ω and ϵ' privileged directions, respectively. These two ray paths, for normal incidence, always lie within the same principal plane (*STU* in Fig. 6–19). This plane also contains (1) the normally incident ray, (2) the crystal's optic axis, and (3) the crystal's ϵ' (or ϵ) privileged direction, these three directions being always coplanar; consequently, this plane is easy to define if two of these three directions are known.

The plane STU, drawn enlarged in Fig. 6–20, is thus a key plane in determining graphically the E-ray path and the precise value of ϵ' for its associated vibration. To do this one must know the actual values of ϵ, ω, and θ—the latter being the angle between the crystal's optic axis and the crystal face upon which normal incidence occurs. Construction details for Fig. 6–20 (and similar graphic solutions) are: (1) Construct an ellipse with ϵ and ω drawn to scale along the semiaxes so as to represent the intersection between plane STU and the indicatrix in Fig. 6–19; (2) through its center and at an angle θ to its ϵ semiaxis (the optic axis of the crystal), draw the line AB, representing the line of intersection in Fig. 6–19 between plane STU and the crystal surface VWX; (3) perpendicular to AB draw the normally incident ray and extend it into the crystal to obtain the O-ray

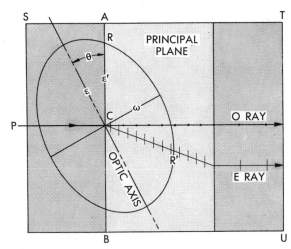

Fig. 6-20. Unobstructed view of plane STU in Fig. 6-19. The cross section between plane STU and the indicatrix produces the ellipse shown, its major axis coinciding with the crystal's optic axis.

path; the O ray's associated vibrations, since they are always perpendicular to the principal plane, are shown as dots; (4) CR, the radius of the ellipse that is parallel to AB, represents both the vibration direction and, by its length, the refractive index ϵ' for the E ray; (5) draw the radius conjugate to CR (*cf.* Appendix I) to obtain CR', the E-ray path.

Other Types of Incidence. Next consider a case like that of ray 2 in Fig. 6–14, wherein the incident ray lies within a principal plane but is not perpendicular to the crystal face. A cross-sectional view along this principal plane (Fig. 6–21) shows the incident ray RC to be split into an

O and an E ray. The angle of refraction for the O-ray path (that is, r_o) is readily determined from Snell's law; thus

$$\sin r_o = \frac{n_i}{\omega} \sin i$$

The angle of refraction of the E-wave normal (that is, r_E) is determinable by a method of successive approximations, each step yielding a value $(r_{E_1}, r_{E_2}, \cdots, r_{E_n})$ more accurate than its predecessor. The procedure, illustrated in Fig. 6–21, is: (1) Assume that the already located O-ray direction (CO) also represents the E-wave normal (it actually does not); (2) draw a perpendicular to CO at C to determine ϵ_1', the refractive index (and vibration direction) associable with such an E-wave normal; (3) substitute this value, ϵ_1', into Snell's law to obtain r_{E_1}, a closer approximation to the angle of refraction for the E-wave normal; thus

$$\sin r_{E_1} = \frac{n_i}{\epsilon_1'} \sin i$$

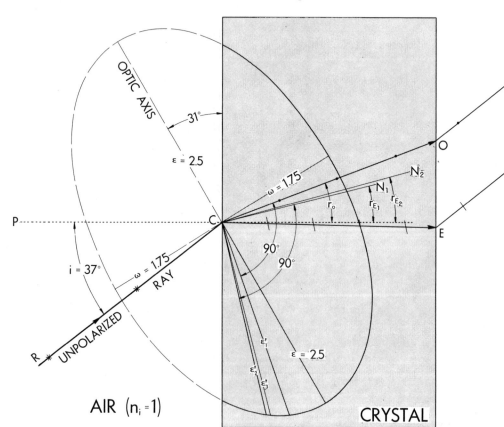

Fig. 6-21. Method of determining refracted ray paths if the unpolarized ray in Fig. 6-19 were obliquely incident (but still in plane STU). All ray paths again lie within plane STU, which is shown above in detail.

(4) using r_{E_1} thus obtained, plot the more accurate E-wave normal direction, CN_1; (5) the perpendicular to CN_1 at point C graphically determines ϵ_2', the refractive index (and vibration direction) associated with CN_1; (6) sub-stitute this value, ϵ_2', into Snell's law to obtain r_{E_2}, an even more accurate angle of refraction for the E-wave normal than r_{E_1}, thus

$$\sin r_{E_2} = \frac{n_i}{\epsilon_2'} \sin i$$

(7) the wave normal thus located (CN_2) is more accurate than CN_1. Thus the perpendicular to CN_2 at point C graphically determines ϵ_3', a refractive index (and vibration direction) that is very close to the true value for the refracted E wave. (8) the E-ray path is then the conjugate radius to vibration direction ϵ_3'.

Fig. 6–21 can also serve as a concrete example. Assume that the crystal plate was cut at a 31-degree angle to its optic axis and that the ray IC is incident upon this plate at a 37-degree angle. If the plate is in air—that is, $n_i = 1.0$—and ω and ϵ have the values shown in the figure, the numerical values obtained for the steps enumerated in the preceding paragraph are: (1) $r_o = 20°07'$; (2) $\epsilon_1' = 2.445$; (3) $r_{E_1} = 14°12'$; (4) $\epsilon_2' = 2.398$; (5) $r_{E_2} = 14°32'$; and (6) $\epsilon_3' = 2.401$. Further substitution of ϵ_3' into Snell's law yields $r_{E_3} = 14°31'$, a value not appreciably different from r_{E_2}. In general, r_{E_2} is sufficiently accurate for most purposes. The reader is invited to check the method using as large a graphic scale as possible.

The most generalized case of ray incidence, wherein the plane of incidence is at a random angle to the optic axis (for example, ray 3 in Fig. 6–14), is too difficult and cumbersome to discuss. However, if the angle between plane of incidence and optic axis is 90 degrees, this case reduces to a very simple one, as illustrated in Fig. 6–22. Here an incident, unpolarized ray IO is resolved after entering the crystal into an O and an E ray. The O-ray path is easily obtained since Snell's law can be solved to locate it; thus

$$n_i \sin i = \omega \sin r_O \qquad \text{(Eq. 6–5)}$$

The O ray's vibration direction is defined since (1) it must be perpendicular to the optic axis (and therefore must lie in the shaded plane of incidence in Fig. 6–22) and (2) it must also be perpendicular to the O-ray path within this plane. In common with the O-ray path, the E-ray path must also lie in the plane of incidence (as all refracted ray paths must). Its vibration direction coincides with the optic axis direction and its corresponding refractive index is therefore ϵ. This vibration direction (drawn on the plate's upper surface) is thus perpendicular to the plane of incidence and consequently to the E-ray path as well. Since the light composing this E ray vibrates perpendicular to the E-ray path, this E ray behaves unusually (for an E ray) in that its path follows Snell's law (cf. pp. 72-3). Thus we can determine r_E, the angle of refraction for both its path and wave normal from a simple Snell's law equation

$$n_i \sin i = \epsilon \sin r_E \qquad \text{(Eq. 6–6)}$$

Since the right sides of Eqs. 6–5 and 6–6 may be equated, we obtain

$$\frac{\omega}{\epsilon} = \frac{\sin r_E}{\sin r_O}$$

For uniaxial negative crystals ($\omega > \epsilon$), therefore, $\angle r_E$ exceeds $\angle r_O$, as is the case for Fig. 6–22. If the crystal in Fig. 6–22 were uniaxial positive ($\epsilon > \omega$), then $\angle r_O$ would exceed $\angle r_E$.

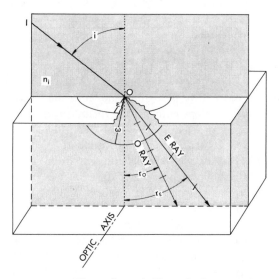

Fig. 6-22. The refracted E- and O-ray paths when an unpolarized ray, IO, is incident within a plane (shaded) perpendicular to the crystal's optic axis. The crystal is uniaxial negative in sign. For a uniaxial positive crystal, r_E would be less than r_O, and the E ray would then make the smaller angle with the perpendicular to the crystal face (dotted line).

SUMMARY

1. The optical indicatrix indicates how refractive index changes with vibration direction in a crystal.

2. The isotropic indicatrix is spherical, indicating that isotropic materials exhibit the same refractive index, regardless of vibration direction of light.

3. For uniaxial crystals the indicatrix is either a prolate spheroid (positive crystals) or an oblate spheroid (negative crystals).

4. All central sections through the indicatrix, except one, are ellipses.

5. The one exception is the circular section.

6. The direction normal to the circular section is called the optic axis. This optic axis always coincides with the c axis of hexagonal and tetragonal crystals.

7. A principal section through the indicatrix is one that contains the optic axis. The ellipse thus formed will always have semiaxes equal to ϵ and ω.

8. The O ray always vibrates *perpendicular* to the principal plane containing its path.

9. The E ray always vibrates *within* the principal plane containing its path.

10. Within this principal plane, the E ray's path is not necessarily perpendicular to its vibration direction. Instead its path and vibration directions are conjugate radii of the ellipse formed by the intersection of this principal plane with the uniaxial indicatrix.

11. A wave normal direction is always perpendicular to its associated vibration directions.

12. The O-wave normal direction and the O-ray path always coincide.

13. The E-wave normal and E-ray path generally do not coincide.

14. The two possible vibration directions associable with a given wave normal are at right angles to it and to each other.

15. Given a wave normal, its associated vibration directions lie within the plane perpendicular to it. More particularly, this plane generally intersects the indicatrix in an ellipse whose major and minor axes represent the only two privileged directions (that is, possible vibration directions) that may be associated with the given wave normal.

16. If this plane intersects the indicatrix in a circle, no privileged directions exist with respect to the given wave normal.

17. For anisotropic crystals, Snell's law must be redefined to apply to wave normal directions rather than ray paths.

18. For normally incident light entering a uniaxial crystal plate from air, the vibration directions are parallel to the crystal boundary both before and after entry (*cf.* revised Snell's law).

19. More specifically, the vibrations after entry are parallel to either or both of the semiaxes of the ellipse which is formed by the intersection of the crystal's boundary plane with the uniaxial indicatrix, the boundary plane extending outward from the indicatrix center.

SEVEN ── THE INTERFERENCE OF LIGHT

WAVES POLARIZED IN THE SAME PLANE

We have described light waves as being transverse waves (see p. 4). Thus light energy streams along its path in a given medium by means of vibrations of the particles of this medium, the vibrations being perpendicular to the path of energy travel. Fig. 7–1A represents, in essence, a motion-stopping photograph of a plane-polarized wave traveling along OP; the arrows transverse to OP represent the vibrational displacements (from their original rest positions on OP) of a few of the particles taking active part in light transmission. Particles on the same wave train are "in phase" if, at any and all given instants, they are always displaced from their rest position by the same direction and amount. Thus, in Fig. 7–1A, points a_1, a_2, and a_3 are all in phase; so are points b_1, b_2, and b_3. Particles such as a_1 and b_1, whose vibrational displacements are equal but opposite in direction, are precisely "out of phase." Points b_1 and c_1 are not in phase since, at a slightly later instant of time, their displacements would be unequal.

The distance between two points on the same wave path is known as their "path difference" and is symbolized as Δ. Since it is a distance, it may be expressed in any convenient units of length; often, however, it is convenient to express path difference in terms of the number of wavelengths of the light being used. We then find that the path difference between all points in phase (for example, a_1 and a_3 in Fig. 7–1A) is always an integral number of wavelengths—that is, 0, 1λ, 2λ, $\cdots$, or $n\lambda$—whereas between all points precisely out of phase (for example, a_1 and b_2) it is $\frac{1}{2}\lambda$, $\frac{3}{2}\lambda$, $\cdots$, or $\frac{2n+1}{2}\lambda$. The path differences between points neither precisely in nor precisely out of phase will be other than these values. Thus, the path difference between two points, if expressed in wavelengths, discloses the extent to which these points are in phase (or out of phase).

In Fig. 7–1B, a "photograph" taken an instant later than Fig. 7–1A, the original wave (amplitude r_1) has traveled farther to the right and a second wave, of equal wavelength but of amplitude r_2, has entered from the left. The path difference between the forefronts of these two waves is now $3\frac{5}{8}\lambda$. The two waves are thus more nearly out of phase than in phase (since $3\frac{5}{8}$ is closer to $3\frac{1}{2}$ than to 4). These two wave motions, simultaneously traveling the same path, do not remain separate entities but rather, as shown in the rectangular inset, combine (that is, interfere) to produce a composite

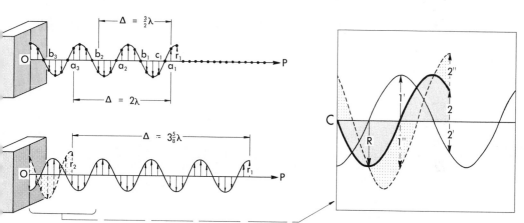

Fig. 7-1. (A) Transmission of a plane-polarized light wave of amplitude r_1 from O toward P by means of transverse displacement of particles (black dots) from their rest positions on OP. (B) A similar "photograph" taken an instant later. A second wave motion of amplitude r_2 has entered from the left and is also traveling from O toward P. (C) An enlarged portion of (B) to show in detail the interaction of these two coinciding wave motions to produce a resultant wave motion of amplitude R.

wave motion (shaded gray) whose amplitude is R. Thus the vibration vectors of the two parent waves, where they coincide in space and time, combine vectorially to produce the vibration vectors of the composite wave. For example, the sum of vectors 1′ plus 1″ is zero whereas that of 2′ and 2″ is vector 2. The form of such composite wave motions depends upon the wavelengths, amplitudes, and path difference of the interfering parent waves. In the common case for optical mineralogy wherein a path difference equal to $x\lambda$ exists between two parent waves of equal wavelengths, the relation between their amplitudes (r_1 and r_2) and the amplitude of the resultant wave (R) is

$$R^2 = r_1^2 + r_2^2 + 2r_1r_2 \cos (x \cdot 360°) \qquad \text{(Eq. 7–1)}$$

Note that if the two interfering waves are in phase—that is, their path difference is 0, 1λ, 2λ, ⋯, or $n\lambda$—Eq. 7–1 indicates R to equal the sum of r_1 and r_2. In this event *constructive interference* occurs, the resultant wave being of greater amplitude than either parent. If the two interfering waves are exactly out of phase—that is, their path difference is $\frac{1}{2}\lambda$, $\frac{3}{2}\lambda$, ⋯, or $\frac{2n + 1}{2}\lambda$—Eq. 7–1 indicates R to equal the difference between r_1 and r_2.

In this event, *destructive interference* occurs, the resultant wave being of lesser amplitude than one or both parents. For such a situation, if r_1 should equal r_2, *total destructive interference* would occur, the two wave motions mutually annihilating each other to produce quiescence rather than a wave motion.

WAVES POLARIZED IN PERPENDICULAR PLANES

For .two waves simultaneously traveling along the same path but vibrating within two mutually perpendicular planes (Fig. 7–2), the resultant wave again represents the vector sum of the vibrations of the two parent waves for all points along the path. In the case illustrated, wherein the points of zero displacement from the path coincide for both waves, the resultant wave (dark gray in Fig. 7–2) will vibrate in a plane that is at an angle to the plane of vibration of either parent. The vectorial addition at a representative point of the two parent vibrations (hollow arrows) to yield the resultant vibration (solid black arrow) is shown on the imaginary plane in Fig. 7–2.

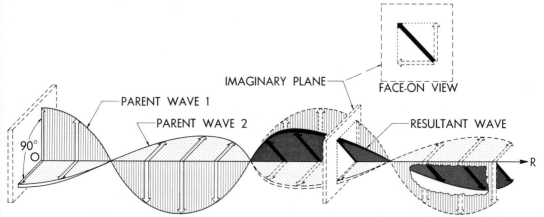

Fig. 7-2. Interference between two waves, polarized in mutually perpendicular planes, to produce a resultant wave motion (polarized in another plane). The imaginary plane indicates how, at any point along *OR*, the vibration vectors of the two parent motions (hollow arrows) may be added vectorially to produce the vibration vectors (solid arrows) of the resultant wave.

Calculation of Path Difference or Retardation

Two perpendicularly plane-polarized wave motions like those in Fig. 7–2 may be considered to emerge from an anisotropic crystal plate upon which plane-polarized light is incident (Fig. 7–3). As illustrated, this plane-polarized light is resolved immediately upon entering at point *O* into a slow wave (for which the crystal's privileged direction and index are ON and N, respectively) and a fast wave (privileged direction and index, On and n).* At this precise instant the two waves are exactly in phase, no path difference existing between them. However, to pass through the thickness of the crystal, t, the slow wave requires the time T_N whereas the fast wave requires only T_n, a lesser amount. Consequently, while "waiting" for the slow wave to emerge, the fast wave travels the distance $c(T_N - T_n)$ in air, c being the velocity of light; Fig. 7–3, essentially a "photograph" taken at

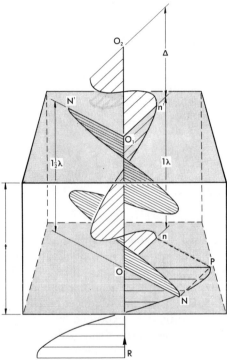

Fig. 7-3. Passage of light through a crystal plate that produces a retardation (Δ) of 270 mμ. If illuminated by light of wavelength 540 mμ, this distance represents $\frac{1}{2}\lambda$. Note that, if a path difference of $\frac{1}{2}\lambda$—or $\frac{2n+1}{2}\lambda$—exists, either ON or On, the light's two vibration vectors upon entry, becomes reversed in direction at its point of exit from the crystal's upper surface (*cf.* ON and ON').

---

* The symbols n and N (respectively read "small en" and "large en") will be used here to refer to the smaller and larger refractive indices of any birefringent plate. For the beginning student this appears to be mnemonically superior to the standard practice of using n_1 and n_2 for these indices.

the precise moment when the slow wave is about to emerge into air, shows the fast wave to have emerged by this distance. This distance is the path difference (Δ) between these two perpendicularly polarized wave trains as they travel in air along direction O_1O_2; thus

$$\Delta = c(T_N - T_n)$$
$$= cT_N - cT_n \qquad \text{(Eq. 7–2)}$$

The actual velocities of the slow and fast waves while traveling in the crystal (here symbolized as c_N and c_n, respectively) are, from the familiar distance-per-time-traveled formula,

$$c_N = \frac{t}{T_N}; \qquad c_n = \frac{t}{T_n} \qquad \text{(Eq. 7–3)}$$

Solving for T_N and T_n, then substituting the values thus obtained into Eq. 7–2, the result is

$$\Delta = \frac{ct}{c_N} - \frac{ct}{c_n}$$
$$= t\left(\frac{c}{c_N} - \frac{c}{c_n}\right) \qquad \text{(Eq. 7–4)}$$

From the definition of refractive index (p. 6), Eq. 7–4 becomes

$$\Delta = t(N - n) \qquad \text{(Eq. 7–5)}$$

Eq. 7–5 indicates that the path difference—that is, retardation of the slow wave behind the fast—is proportional to (1) the plate's thickness and (2) the difference in refractive indices of its two privileged directions.

The term $(N - n)$ is dimensionless; hence path difference, if computed from Eq. 7–5, is expressed in whatever units of distance are used for t. Path difference is traditionally expressed in millimicrons (mμ)* whereas the crystal thicknesses ordinarily dealt with in optical mineralogy are usually expressed in millimeters (mm). Thus it is necessary to convert the crystal's thickness from mm to mμ by means of the conversion factor

$$1 \text{ mm} = 10^6 \text{ m}\mu$$

For example, assume a crystal plate 0.03 mm in thickness to possess refractive indices of 1.553 and 1.544 for its two privileged directions.. For this plate

$$\Delta = (1.553 - 1.544).03 \text{ mm}$$
$$= 0.009 \times .030 \times 10^6 \text{ m}\mu$$
$$= 270 \text{ m}\mu$$

---

* Prior to this chapter the wavelengths of light have been expressed in Angstrom units to conform to an almost universal practice of physicists. In this chapter, however, wavelength will be expressed in mμ, the unit of length commonly used for path differences. For example, since 1 mμ equals 10 Å, the wavelength of sodium light will now be expressed as 589.3 mμ.

From the discussion on page 95 the reader was led to infer that, after their emergence from the crystal in Fig. 7–3, the slow and fast wave trains simultaneously traveled upward along O_1O_2 as two discrete wave motions. Actually, however, as discussed on page 94, these two waves mutually interfere to produce a resultant wave motion. Depending upon the path difference between these two interfering wave trains, this resultant wave motion may be plane polarized, circularly polarized, or elliptically polarized. The nature of these types of polarization and their relation to the path difference between the two interfering waves will be discussed next.

Plane Polarization of the Resultant Wave

Assume that the slow wave required two periods and the fast wave only one to pass through the crystal of Fig. 7–4A.* After their emergence, therefore, a path difference of λ exists between the two wave trains. Consequently, as is always the case for path differences of 0λ, 1λ, 2λ, $\cdots$, $n\lambda$, the two waves emergent at O_1 are in phase; that is, at any point along O_1O_2 the vibrations of these two waves are either (1) both in the same direction as On and ON or (2) both opposite, On and ON representing the vibrations of the two waves immediately after they entered the crystal at O (when they were exactly in phase).

Within the crystal the two wave trains were incoherent—that is, incapable of mutually interfering; after their emergence, however, they become coherent. Thus, instead of proceeding upward as the two discrete wave motions indicated by the dashed curves in Fig. 7–4A, their vibration vectors are vectorially additive at all points of coincidence in space and time, points O_1, a, and b being detailed examples. As shown, their resultant wave is plane polarized and vibrates in precisely the same plane as did the light incident on the crystal (as is always the case for path differences of 0, 1λ, 2λ, $\cdots$, $n\lambda$). Note that, if the analyzer is inserted with its privileged direction perpendicular to that of the polarizer (that is, the nicols are crossed), the analyzer will not transmit any of this light. If, however, its privileged direction is parallel to the polarizer's, it will transmit all of this light.

Next assume the slow wave to require two periods and the fast wave one and one-half to pass through the crystal (Fig. 7–4B). Upon their emergence, a path difference of $\frac{1}{2}\lambda$ exists between these waves. Consequently, at O_1 one of these waves (but never both) will always be vibrating opposite in direction from On or ON, as the case may be. The mutual interference of these two wave motions (shown in detail at points O_1, a, and b) thus produces a plane-polarized resultant whose plane is at 90 degrees to that of the incident light. Thus, the light from this crystal, if viewed between

---

* A period is the time required for the series of vibrations necessary to complete one full wavelength. As shown in Fig. 7–4A, therefore, the slow wave traced out two wavelengths, the fast wave only one, while passing through the crystal. The period of a wave motion is technically defined as the reciprocal of the frequency and is constant for a given, highly pure color of light.

crossed nicols, will be fully transmitted by the analyzer; if viewed between parallel nicols, none will be transmitted. Similar results would be obtained for crystals that produced path differences of $\frac{3}{2}\lambda$, $\frac{5}{2}\lambda$, $\cdots$, $\frac{2n+1}{2}\lambda$.

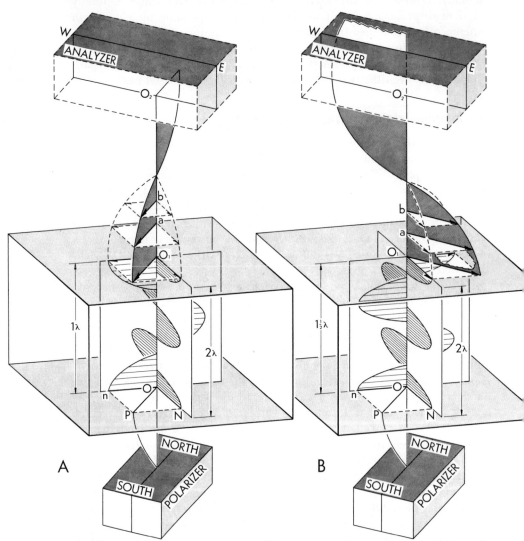

Fig. 7-4. The mutual interference, after their emergence from an anisotropic crystal at 45 degrees off extinction, of what were formerly the slow and fast waves within the crystal. In (A), passage through the crystal has produced a path difference of 1λ between these waves, and their resultant wave motion after exit at O_1 is polarized in the same plane as the light from the polarizer. For (B), where their path difference is $\frac{1}{2}\lambda$, the resultant wave motion after exit at O_1 is at 90 degrees to the plane of polarization of light from the polarizer. Between crossed nicols, therefore, the resultant light emergent from the crystal is for (A) completely extinguished by the analyzer but for (B) completely transmitted (reflectional and absorptional losses being neglected).

Circular Polarization

If the path difference between the two waves emergent from a crystal is $\frac{1}{4}\lambda$ (or $\frac{3}{4}\lambda$, $\frac{5}{4}\lambda$, $\cdots$, $\frac{2n+1}{4}\lambda$), their vibrations interfere, where they coincide in space and time, to produce resultant vibration vectors of constant lengths but variable azimuths (arrows in Fig. 7–5A). The resultant wave

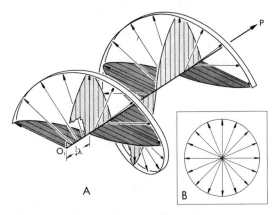

Fig. 7-5. (A) Production of circularly polarized light by the interaction of two, mutually perpendicular, plane-polarized waves whose path difference is $\frac{1}{4}\lambda$; O_1 represents their point of emergence from a crystal. These wave motions interact to produce spirally distributed vibrations, only a few of which are shown. A ribbon tangent to these arrow tips has been added to emphasize the spiral nature of the surface joining all these resultant vibrations. A perpendicular dropped from any point on this ribbon to O_1P indicates a vibration direction. (B) Outline of the spiral surface joining these vibration directions as seen looking down O_1P.

motion thus consists of vibration vectors that spiral upward and define a surface* resembling the thread of a screw. If it could be viewed by looking down O_1P (the direction this wave motion is traveling), this surface would appear circular in outline (Fig. 7–5B); hence this light is said to be circularly polarized.

Elliptical Polarization

The resultant wave motion emergent from crystals that produce path differences of other than 0, $\frac{1}{4}\lambda$, $\frac{1}{2}\lambda$, $\frac{3}{4}\lambda$, 1λ, $\cdots$, $\frac{n+1}{4}\lambda$ spirals upward as for circular polarization. Now, however, the vibration vectors no longer maintain a constancy of length. Thus, if viewed as for Fig. 7–5B, the spiral surface connecting these vibration vectors is elliptical rather than circular

---

* Known in mathematics as a Riemann surface.

in outline; hence the light is said to be elliptically polarized. Elliptical polarization is the most frequent and general case, circular and plane polarization often being regarded as special cases.

Common Conventions

Any plane, circular, or elliptically polarized wave may thus represent the interaction between two, mutually perpendicular, plane-polarized wave motions. Conversely, just as a vector may be resolved into and be replaced by two component vectors, so may any planar, circular, or elliptical wave motion be resolved into and replaced (in our thoughts) by two, mutually perpendicular, plane-polarized waves. In optical mineralogy it is sometimes convenient to assume that the slow and fast waves do not interact after their emergence from the crystal, thus avoiding the necessity of drawing elliptical polarization as well as simplifying the discussion.

TRANSMISSION BY THE ANALYZER

After its successive passage through (1) the polarizer, (2) a crystal, and (3) the analyzer, the percentage of a monochromatic beam transmitted by the analyzer depends upon (a) ϕ, the angle between the privileged directions of the polarizer and the analyzer (see p. 70); (b) τ, the angle between the polarizer's privileged direction and the crystal's closest privileged direction (see p. 85); (c) Δ, the path difference produced between the slow and fast waves during their transmission through the crystal; and (d) the wavelength (λ) of the light. Following Johannsen (1918, p. 343 ff.), the general relationship is

$$L = 100\left[\cos^2\phi - \sin 2(\tau - \phi) \cdot \sin 2\,\tau \cdot \sin^2\left(\frac{\Delta}{\lambda}\right) 180°\right] \quad \text{(Eq. 7–6)}$$

where L represents the percentage of transmission by the analyzer—that is, the ratio (converted to percent) of the intensity of the light immediately before and immediately after transmission by the analyzer. Eq. 7–6 assumes there is no loss of light through reflection or absorption during passage through the crystal or lens system of the microscope.

For the important case in which a crystal is viewed at 45 degrees off extinction between crossed nicols (that is, τ equals 45 degrees and ϕ equals 90 degrees), Eq. 7–6 reduces to

$$L = 100 \sin^2 \frac{\Delta}{\lambda} 180° \quad\quad\quad \text{(Eq. 7–7)}$$

If the crystal is viewed in this same position between parallel nicols (that is, τ equals 45 degrees and ϕ equals 0 degrees), Eq. 7–6 becomes

$$L = 100 \cos^2 \frac{\Delta}{\lambda} 180° \quad\quad\quad \text{(Eq. 7–8)}$$

Fig. 7–6, based on Eqs. 7–7 and 7–8, graphically illustrates how L, the percent transmission by the analyzer, varies for a crystal at 45 degrees off extinction according to this crystal's path difference (or phase difference). The left-hand scale cites percentages if the nicols are crossed; the right-hand scale, those for parallel nicols.

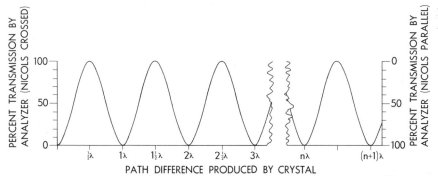

PATH DIFFERENCE PRODUCED BY CRYSTAL

Fig. 7-6. Transmission of light by the analyzer according to the phase difference produced in this light by prior passage through a crystal (at 45 degrees off extinction). Use left-hand scale if nicols are crossed; right-hand scale if they are parallel. The 100 percent value signifies that all the light incident on the crystal from the polarizer also passes through the analyzer (all losses due to reflection and absorption being neglected).

INTERFERENCE COLORS

Origin

If illuminated with white light, plates of anisotropic crystals viewed (off extinction) between crossed or parallel nicols will appear colored. These colors, known as interference colors, result from the unequal transmission by the analyzer of the component wavelengths of the white light. The particular wavelengths that the analyzer transmits (or absorbs) depend upon the amount of retardation produced in the light by its prior passage through the crystal. Particular values of retardation produce particular interference colors. For example, suppose that a crystal whose retardation is 550 mμ is viewed at 45 degrees off extinction while illuminated by white light. With respect to the 400.0, 440.0, 488.9, 550.0, 628.6, and 733.3-mμ wavelength components of this white light, to cite just a few, this 550-mμ retardation value respectively represents a path difference of $1\frac{3}{8}\lambda$ (that is, 550/400), $1\frac{1}{4}\lambda$ (that is, 550/440), $1\frac{1}{8}\lambda$ (that is, 550/488.9), 1λ (that is, 550/550), $\frac{7}{8}\lambda$ (that is, 550/628.6), and $\frac{3}{4}\lambda$ (that is, 550/733.3). The percentage of each of these wavelengths which is transmitted by the analyzer for either crossed or parallel nicols may therefore be quickly determined from Fig. 7–6. A plot of the values thus obtained (curve A in Fig. 7–7) indicates that only a little of the green and yellow wavelength is transmitted by the analyzer if this crystal is viewed between crossed nicols, and consequently it appears reddish violet in color; viewed between parallel nicols, the same crystal (now refer curve A to the right-hand scale in Fig. 7–7) would more abundantly transmit green light and thus appear green. The transmission curve for the analyzer may be similarly determined for a crystal of 800-mμ retardation (curve B in Fig. 7–7) viewed between crossed or parallel nicols.

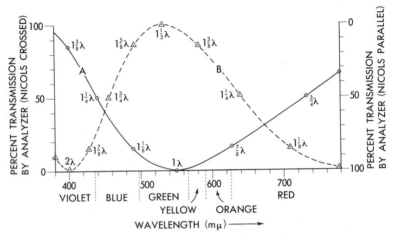

Fig. 7-7. Percent transmission by the analyzer of the different wavelengths in a beam of white light that has first passed through a crystal of retardation 550 mμ (curve A) or of retardation 800 mμ (curve B), both crystals being at 45 degrees off extinction. For a crystal viewed between crossed nicols, read the left-hand scale; for parallel nicols, read the right-hand scale. In case A, therefore, the crystal appears reddish violet between crossed nicols but green between parallel nicols. In case B this is reversed. The path differences developed for a few specific wavelengths of light (by their passage through each crystal) are shown at several points on both curves. For example, light of wavelength 400 mμ in the incident white light develops a path difference of 2λ during passage through the crystal associated with curve B but a path difference of $1^{3}/_{8}\,\lambda$ during passage through the crystal associated with curve A.

Classification and Nomenclature

The relationships between particular retardation values and the interference colors characteristic of them for crossed nicols are summarized in Fig. 8–17 (fol. p. 144), where a vertical bar of the color appears above the retardation value that produces it. In the figure these interference colors are divided into orders according to whether they result from retardations of 0 to 550 mμ (first-order colors), 550 to 1100 mμ (second-order colors), 1100 to 1650 mμ (third-order colors), and so on. A red interference color, for example, if it results from a retardation of 1650 mμ

is called third-order red, written more succinctly as 3° red.* A red caused by a retardation of 550 mμ would thus be written 1° red (and called first-order red). The interference color corresponding to 560 mμ, which begins the second order, is readily discriminated by the human eye from the interference colors associable with retardations slightly more or less than 560 mμ. Consequently this color is called the *sensitive tint, sensitive violet,* or "tint of passage." The alternating pink and green interference colors associated with retardations increasingly greater than 2300 mμ become milkier and milkier until, for very high retardations regardless of their value, the interference color is always white. This white is called a *high-order white* to distinguish it from first-order white, which is associated with a retardation of about 200 mμ. The analyzer's transmission curve when first-order white is being viewed (dashed in Fig. 7–8) differs strikingly from that obtained when a typical high-order white is being viewed (solid line). Although, superficially, both colors look alike between crossed nicols,

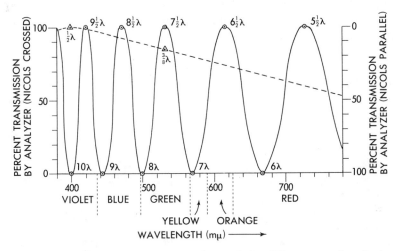

Fig. 7-8. Transmission by the analyzer of the different wavelengths in a beam of white light that has first passed through a crystal (at 45 degrees off extinction) of retardation 200 mμ (dashed line) or of retardation 4000 mμ (solid line). The light thus transmitted (for crossed nicols) is interpreted by the eye as white. The dashed transmission curve represents the first-order white of Fig. 8-17; the sinuous curve, a "high-order white." The path differences developed for a few specific wavelengths of light (by their passage through each crystal) are shown at several points on both curves. Note that, if the analyzer's vibration direction were rotated into parallel position with the polarizer, the high-order white would remain white whereas the first-order white observed would give way to a reddish color.

---

* The symbols 1°, 2°, 3°, 4°, ···, will henceforward represent the terms "first order," "second order," "third order," "fourth order," ··· . This convention is particularly convenient in illustrations where space is limited. Confusion with the universal use of the degree symbol for temperature or angular measurement is unlikely.

with experience they can be distinguished at a glance since high-order white appears more creamy, whereas first-order white appears more bluish. They are further distinguishable through insertion of an accessory (see p. 124 ff.), since it will produce a pronounced change in color for first-order white but not for a high-order white.

Effect of Rotation

Observed during rotation of the stage, the interference color of an anisotropic plate changes in intensity but not in hue. To understand this, consider first those wavelengths of the white light passing through the crystal whose slow waves undergo retardations equivalent to path differences of 1λ, 2λ, 3λ, $\cdots$, or $n\lambda$. Upon emergence from the crystal at 45 degrees off extinction, these wavelengths form resultant wave motions that are plane polarized parallel to the incident light (Fig. 7–4A illustrates this in detail). In Fig. 7–9A this is shown less elaborately for a cylinder cut from the crystal of Fig. 7–4A, the sinusoidal outlines of the wave motions having been omitted. Symbolism is the same as for Fig. 7–4A: OP represents the vibration direction and amplitude of the light from the polarizer; On and ON, the vibration directions and amplitude of the two light waves into which the incident light is resolved immediately upon entry into the crystal. To indicate in Fig. 7–9A that a path difference of 1λ, 2λ, 3λ $\cdots$, or $n\lambda$ exists between the two waves emerging at O_1, O_1n_1 and O_1N_1 are drawn to

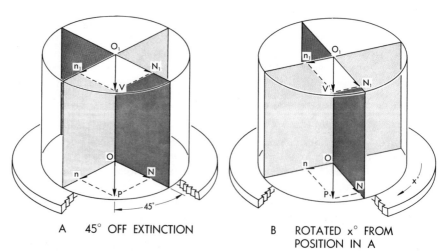

| A   45° OFF EXTINCTION | B   ROTATED x° FROM POSITION IN A |

Fig. 7-9A-B. Effect of rotation on an interference color. (A) Crystal at 45 degrees off extinction; light undergoing path difference of 1λ, 2λ, ... or $n\lambda$ during passage through crystal cylinder. The resultant wave motion emerging from the cylinder vibrates parallel to O_1V, whose direction coincides with OP, the polarizer's privileged direction. (B) Crystal rotated x degrees clockwise from position in (A); the plane of vibration of the resultant wave motion after this rotation—that is O_1V'—is the same direction as O_1V in (A).

point in the same direction as On and ON. Study of Fig. 7–4A illustrates the logic behind this convention. The resultant of O_1n_1 and O_1N_1 is O_1V. Thus the resultant of the light emerging from the crystal vibrates parallel to the incident light. Rotation of the stage through an angle x degrees clockwise (Fig. 7–9B) changes the directions and lengths of On_1 and ON_1, but not of their resultant O_1V'. Thus, regardless of position of the stage, the resultant light would be completely extinguished by the analyzer (crossed nicols assumed).

Consider next the wavelengths of the white light whose slow waves undergo retardations equivalent to path differences of $\frac{1}{2}\lambda$, $\frac{3}{2}\lambda$, $\cdots$, or $\frac{2n+1}{2}\lambda$. Upon emergence from the crystal at 45 degrees off extinction, these wavelengths form resultant wave motions that vibrate at 90 degrees to the vibration direction of the incident light (see Fig. 7–4B). Fig. 7–9C, a cylinder cut from the crystal of Fig. 7–4B, shows this more simply. To indicate a path difference of $\frac{1}{2}\lambda$, $\frac{3}{2}\lambda$ $\cdots$, or $\frac{2n+1}{2}\lambda$, On_1 is drawn with its direction reversed from On. Study of Fig. 7–4B discloses the logic behind this. Because of this reversal, the resultant of O_1n_1 and O_1N_1 (that is, O_1V) makes a 90 degree angle to OP. Thus, for the crystal at 45 degrees off an extinction position, the light emergent from the crystal vibrates at 90 degrees to what it did prior to entry. This resultant light would therefore be completely transmitted by the analyzer (crossed nicols assumed). As the crystal is rotated x degrees clockwise from its position in Fig. 7–9C—that is, x degrees from its position at 45 degrees off extinction—the resultant vector O_1V' (see Fig. 7–9D) rotates through an angle of $2x$ degrees from its former

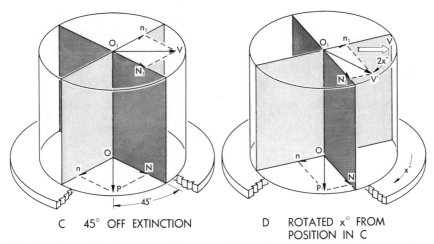

C   45° OFF EXTINCTION   D   ROTATED x° FROM
                                           POSITION IN C

Fig. 7-9C-D. (C) Crystal at 45 degrees off extinction; light undergoing a path difference of $\frac{1}{2}\lambda$, $\frac{3}{2}\lambda$, ... or $\frac{2n+1}{2}\lambda$ during passage through crystal cylinder. The resultant wave motion emerging from the cylinder vibrates parallel to O_1V which is now at 90 degrees to OP, the polarizer's privileged direction. (D) Crystal rotated x degrees clockwise from its position in (C). The plane of vibration of the resultant wave motion after this rotation—that is, O_1V'—now vibrates at $(90° - 2x°)$ to OP, the polarizer's privileged direction.

direction (O_1V). Since O_1V' is no longer parallel to the analyzer's privileged direction (for crossed nicols), the resultant light is no longer completely transmitted by the analyzer. As the crystal is rotated toward an extinction position, the resultant O_1V' becomes increasingly parallel to OP. .

Thus wavelengths of light that undergo a path difference of 1λ, 2λ, $\cdots$, $n\lambda$ "suffer" total extinction at the analyzer regardless of the position of the stage, whereas wavelengths which undergo phase differences of $\frac{1}{2}\lambda$, $\frac{3}{2}\lambda$, $\cdots$, $\frac{2n+1}{2}\lambda$ are completely transmitted at 45 degrees off extinction. However, less and less of the latter wavelengths are transmitted by the analyzer as the crystal is turned toward an extinction position. Consequently, if an anisotropic crystal plate is observed between crossed nicols during a 360-degree rotation of the stage, there will be four positions (representing 45 degrees off extinction) for which the interference color will appear brightest. Similarly there will be four "extinction positions" (at which the plate's privileged directions will be parallel to those of the polarizer and analyzer) for which the interference color will be at zero intensity. The existence of interference colors and the attendant extinction positions as a grain is rotated is a criterion for anisotropy in crystals.

ORTHOSCOPIC AND CONOSCOPIC OBSERVATION OF INTERFERENCE EFFECTS

In optical mineralogy the interference phenomena produced by the action of anisotropic crystals on light may be observed with the microscope arranged as (1) an orthoscope or (2) a conoscope. The orthoscopic arrangement may be regarded as a normal microscope arrangement in which are inserted the polarizer (below the stage) and the analyzer (above the stage), their privileged directions being mutually perpendicular. The conoscopic arrangement requires, in addition to these, insertion of a Bertrand lens and

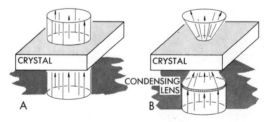

Fig. 7-10. Comparison of the ray paths for orthoscopic illumination (A) and conoscopic illumination (B). For the actual situation the rays shown in (A) would be slightly convergent. They become increasingly parallel, however, as the apertures are decreased (for example, by closing down the iris diaphragm or inserting a lower power lens).

a substage condensing lens. The latter causes the object on the stage to be illuminated by a cone of light (Fig. 7–10B) rather than by a bundle of near-parallel rays as it is with the orthoscope (Fig. 7–10A). For the orthoscope, therefore, the crystal is illuminated by a series of essentially parallel, normally incident rays all of which travel along the same crystallographic direction within the crystal. For the conoscope, only the central ray within the illuminating cone is normally incident; moreover, the various rays of the cone travel along different crystallographic directions within the crystal. The differences in illumination for the two methods, coupled with the optical effect produced by insertion of the Bertrand lens (or, alternatively, by removal of the ocular) in the conoscopic arrangement, result in entirely different interference phenomena being observed by the two methods.* Table 7–1 indicates the microscope set-ups suggested for the two methods.

Table 7–1

Suggested microscope arrangements

		Orthoscope	Conoscope
Bertrand lens		Not inserted	Inserted.[1] If equipped with iris diaphragm, its aperture may be reduced to sharpen figure (especially for small crystals)
Analyzer		Inserted (Privileged direction at right angle to that of polarizer)	
Objectives		Low, medium, or high according to magnification desired	High-power objective only (N. A. 0.85 preferred)
Substage assembly	Condensing lens	Swing-out lens not inserted	Swing-out lens inserted (or alternative device used to make cone of light illuminating stage more convergent)
	Iris diaphragm	Reduced as needed to sharpen detail	Open; later reduce to sharpen detail
	Polarizer	Inserted	

[1] *N.B.* If the microscope lacks a Bertrand lens, the conoscopic optical arrangement can still be obtained by substituting a pin-hole eyepiece for the ocular.

* The different ray paths involved in the orthoscopic and conoscopic arrangements are discussed in detail by Rinne and Berek (1953, pp. 38-40),

ORTHOSCOPIC EXAMINATION OF CRYSTALS

In earlier discussions it was tacitly assumed that the microscope was being used as an orthoscope. In summary, a crystal under orthoscopic examination, if illuminated by white light, will possess an interference color whose hue is dependent upon its retardation—that is, the product of t, its thickness, and $(N - n)$, the difference between the indices for its privileged directions. Along any thinner edges of the grain (Fig. 7–11) the retardation will be less and interference colors will be lower than those at its thick central part. All points of equal thickness on the grain will be marked by the same interference color, the lines connecting these points of identical interference colors being called isochromes—that is, lines of equal color. When the privileged directions of the grain (ON and On) are at 45 degrees to those of the polarizers, these interference colors are at maximum brightness. As the crystal is rotated toward an extinction position, the intensities of these colors decrease toward zero and the crystal appears black at the extinction position.

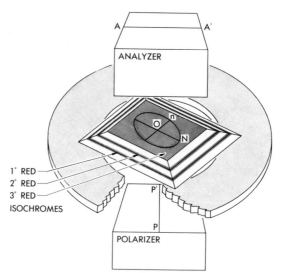

Fig. 7-11. Crystal at 45 degrees off extinction on rotatable stage. Its priviledged directions, ON and On, become parallel to either AA' or PP' four times during a complete rotation of stage. At these times the crystal is at extinction.

CONOSCOPIC EXAMINATION (INTERFERENCE FIGURES)

Viewed with the microscope set up as a conoscope, the images of anisotropic crystals as seen with the orthoscope are supplanted by highly informative patterns of interference colors called *interference figures*. These are formed by rays that traveled along different directions while within the crystal being viewed. Accurate observation of interference figures yields a considerable amount of optic data within a minimum of time. For uniaxial crystals, the interference figure consists of two intersecting black bars—or, as they are called, isogyres—that form a cross (Fig. 7–12) resembling the

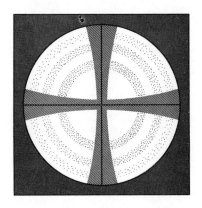

Fig. 7-12. Interference figure of a uniaxial crystal oriented so that its optic axis is perpendicular to the plane of the microscope stage. The common center of the isogyre cross (shaded dark) and the isochrome circles (stippled) represents the point of emergence of rays that, when in the crystal, traveled along the optic axis.

Formée cross of heraldry. This cross is concentric with a series of circles that, if monochromatic light is being used to illuminate the crystal, represent alternations of darkness and brightness for this light. If white light is the illuminant, they represent circular distributions of the interference colors in Fig. 8–17 (follows p. 144), the inner circles being marked by increasingly lower order colors. Since each such circle connects points of identical interference colors, the circles are usually called isochromatic curves or, more briefly, *isochromes*. The common center of both the black cross and isochromes is a black spot (melatope), which marks in the field of view the outcrop of light rays that traveled along the optic axis while in the crystal.

Origin of the Isochromes; Cones of Equal Retardation

Consider the crystal whose interference figure was shown in Fig. 7–12. Rays 1 and 2 in its cross-sectional view (Fig. 7–13A) represent two, almost infinitely close, neighboring rays of the light cone; thus they are essentially parallel.* After entry into the crystal their E and O rays travel separate paths. However, assume that the spacing of rays 1 and 2, which is highly exaggerated in Fig. 7–13A (as is the crystal thickness), is such that the O ray from ray 1 (that is ray 1_O) and the E ray from ray 2 (that is, ray 2_E) travel identical paths after exit from the crystal. They thus interfere according to the path difference produced between them by their passage through the

_____

* Just as two neighboring radii that subtend an infinitely small arc of a circle are parallel.

crystal. In Fig. 7–13A, ray 1_O is 5λ behind ray 2_E after their emergence since, to pass through the crystal, ray 1_O required eight periods but ray 2_E only three.

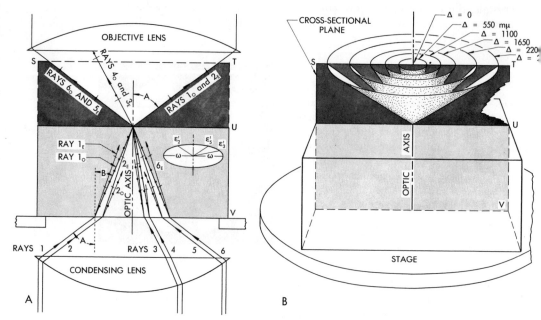

Fig. 7-13. (A) Cross-sectional view of a conoscopically illuminated uniaxial crystal whose optic axis is perpendicular to stage. (B) Cones of equal retardation for same in perspective view. $STUV$ is the cross-sectional plane drawn in (A). Crystal thickness is highly exaggerated.

Similarly, consider two parallel rays that travel at a lesser angle to the optic axis—for example, rays 3 and 4. After exit, rays 4_O and 3_E interfere according to their path difference. Obviously, the path difference between rays 4_O and 3_E is not as great as that between 1_O and 2_E since (1) they did not travel through as large a thickness of crystal and (2) their refractive index difference, $(\omega - \epsilon_3')$, is less than that for rays 1 and 2—that is $(\omega - \epsilon_2')$. Thus emerges the principle that the path difference between the E and O rays after exit is proportional to the angle between their common path and the crystal's optic axis. Therefore, for the rays that traveled parallel to the optic axis both before and after exit (see pp. 81–2), the retardation is zero. The path difference—that is, retardation—between rays 6_O and 5_E, since their angle to the optic axis equals that of 1_O and 2_E, precisely equals the retardation between rays 1_O and 2_E. Considered three dimensionally, all those ray paths that make an equal angle to the optic axis form cones of equal retardation; in Fig. 7–13B only five such cones are drawn, one connecting all coincident E and O ray exit paths between which the retardation is 550 mμ; a second for 1100 mμ; and so on.

The origin of the circular isochromes becomes apparent if one considers an interference figure as bringing into focus the rays traveling along the cones of equal retardation above a crystal. Thus an interference figure may be regarded as a visualization of the outcrops of these cones on a horizontal plane above the crystal.* For example, if the crystal in Fig. 7–13B were illuminated by light of wavelength 550 mμ, the path difference between the E and O rays traveling along the 0, 550, 1100, 1650, 2200, and 2750-mμ retardation cones would be 0λ, 1λ, 2λ, 3λ, 4λ, and 5λ, respectively. Consequently (p. 97), the resultant of these rays is completely extinguished at the analyzer (crossed nicols assumed) and the outcrops of these cones are thus marked by complete extinction. Roughly intermediate between these circles of blackness, however, are the outcrops of retardation cones 275, 825, 1375, 1925, and 2475 mμ (these cones not drawn in Fig. 7–13B). For light of 550-mμ wavelength these cones correspond to path differences of $\frac{1}{2}\lambda$, $1\frac{1}{2}\lambda$, $2\frac{1}{2}\lambda$, $3\frac{1}{2}\lambda$, and $4\frac{1}{2}\lambda$. Hence their outcrops are sites of brightness. Note that all these circles are concentric, the optic axis emerging at their common center. The approximate radii of the circles corresponding to retardations of 2λ, 3λ, 4λ, ⋯, nλ are respectively proportional to $\sqrt{2}\,r$, $\sqrt{3}\,r$, $\sqrt{4}\,r$, ⋯, $\sqrt{n}\,r$, where r is the radius for the circle of retardation 1λ.

If white light is the illuminant, then the outcrop of each cone of equal retardation is marked by a circular isochrome composed of the interference color corresponding to this retardation. In Fig. 7–13B, for example, the 1° red, 2° red, 3° red, 4° red, and 5° red isochromes would respectively mark the outcrops of the 550, 1100, 1650, 2200, and 2750 mμ cones of retardation. Similarly, cones corresponding to values of retardation intermediate between these would give rise to isochromes marked by interference colors (see Fig. 8–17) intermediate between these red colors.

It is sometimes convenient to assume that the E and O rays emerging above a crystal do not interact but instead vibrate as separate entities. Under

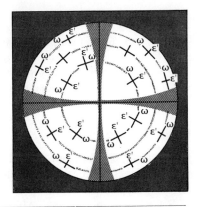

Fig. 7-14. Vibration directions of the E and O rays emerging in the field of view of a uniaxial interference figure, assuming these rays do not interfere after emergence from the crystal.

---

* The advanced student is referred to a technically more accurate statement by Kamb (1958, p. 1032) that interference phenomena essentially occur on the focal sphere of the objective lens; the view of this focal sphere as seen through the ocular, however, is essentially that of this horizontal plane.

this assumption, note from the study of Fig. 7–13 that the E ray vibrates perpendicular to the cones of equal retardation whereas the O ray vibrates tangential to them (but perpendicular to the E ray's vibration). Similarly, under this assumption, all O rays in a uniaxial interference figure (Fig. 7–14) may be considered to vibrate tangential to the isochromes whereas the E rays vibrate parallel to radii of the circular isochromes.

The interference figures for crystals of high birefringence ($\epsilon - \omega$) possess more isochromes than do those for crystals of low birefringence. This also holds true for a thick section as compared to a thinner section of the same mineral (Fig. 7–15). The higher the numerical aperture of the objective used, the wider the angle of the cone of light from the crystal that enters the objective. The cones drawn above the crystal in Fig. 7–15 are those that would enter an objective of N. A. 0.85. If an objective of N. A. 0.65 is used, only the dashed cone of light enters the objective; consequently, only that portion of the interference figure within the dashed circles would be observed. From this standpoint, objectives of N. A. 0.85 are preferable.

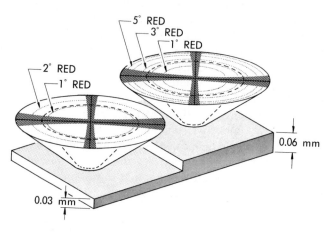

Fig. 7-15. Comparison of the number of isochromes seen in the field of view for two different thicknesses of calcite. If an objective of N.A. 0.65 is substituted for the objective of N. A. 0.85, only the portion of the interference figure within the dashed circles will be seen.

Origin of the Isogyres

A cone of light from the polarizer is composed of rays that, if rotation of their planes of vibration by reflectional losses at lens surfaces is neglected, all vibrate parallel to the privileged direction of the polarizer—that is, N–S for most microscopes. After their passage through a uniaxial crystal as E and O rays and their subsequent interference after exit, the rays in the cone of emerging light now vibrate in diverse directions. Those rays that have retained a N–S or nearly N–S vibration direction will, however, be extinguished or nearly extinguished at the analyzer. The sites of emergence of these extinguished rays in the field of view are thus marked by wedge-shaped areas of extinction commonly called isogyres.

The shape of the isogyres becomes understandable if one determines the vibration directions for the rays emerging at various points in the field of view. To permit greater enlargement, consider only a quadrant of a cone of light emerging from the upper surface of a crystal whose optic axis is perpendicular to the microscope stage (Fig. 7–16A). Within this cone only the 1λ and $1\frac{1}{2}\lambda$ cones of equal retardation are drawn. Let plane OSS' and those at $10°$, $45°$, and $90°$ to it represent extensions above the crystal of a few principal planes, OSS' being parallel to the privileged direction of the polarizer. The intersections of cones and planes mark the paths of representative rays 1 to 8 in the light cone emergent from the crystal. Of these, rays 2, 3, 6, and 7 are formed by the interference of an O and an E ray whose vibration directions were V_O' and V_E' upon exit from the crystal's upper surface. The vibration direction of such resultant rays, as shown for rays 2, 3, and 7 in Fig. 7–16A, is thus V_C, the vector sum of V_O' and V_E'.

The direction of this vector sum, V_C, is shown by double-barbed, solid arrows for rays 1 to 8 at their points of emergence in Fig. 7–16B. Fig. 7–16B may be alternatively regarded as plane SOW in Fig. 7–16A or as a quadrant of a uniaxial interference figure for which the optic axis emerges at O, the cross-hair intersection. The direction of V_C was determined vectorially as follows. At the points of emergence of rays 1 to 8, let the double-barbed, hollow arrows represent the amplitude and vibration direction of rays entering the crystal from the polarizer. Let the single-barbed, hollow arrows, V_O and V_E, represent the same for the O and E rays into which these parent rays are resolved upon entering the crystal. Note that V_E is always parallel and V_O is always perpendicular to the trace of the principal plane (OS, OP, OQ, or OW) in which the ray travels. Single-barbed, solid arrows V_E' and V_O'—which represent the vibration directions and amplitudes of the E and O rays after their passage through the crystal—are drawn reversed to V_E or V_O if, at exit, either ray is out of phase (see p. 97) with what it was upon entry. Thus, on the $1\frac{1}{2}\lambda$ cone, V_E' has been reversed with respect to V_E*. Upon emergence from the crystal, the E and O rays interfere to produce a composite ray whose "vibration vector," V_C, is the vector sum of V_O' and V_E'. For all rays emerging on 1λ (or $n\,\lambda$) retardation cones, note that V_C is N–S; the composite rays traveling along these cones thus vibrate N–S once more. For rays emerging on the $1\frac{1}{2}\lambda$—or $(n + \frac{1}{2})\lambda$—cones, V_C makes an angle $2x$ to the polarizer's privileged direction (OS), where x is defined as the angle between OS and the trace of the principal plane containing the ray. For rays 5, 6, 7, and 8, for example, x respectively equals $0°$, $10°$, $45°$, and $90°$. Thus V_C, their vibration direction after

---

* These conventions were explained during discussions of Figs. 7–4A and 7–4B on pp. 97–98.

emergence from the crystal, is respectively at 0°, 20°, 90° and 180° to OS. The component of V_C that is parallel to $A'A'$, the analyzer's privileged direction—for example, OA as shown for ray 6 in the circular inset in Fig. 7–16—is a measure of the degree to which the ray is transmitted by the analyzer.

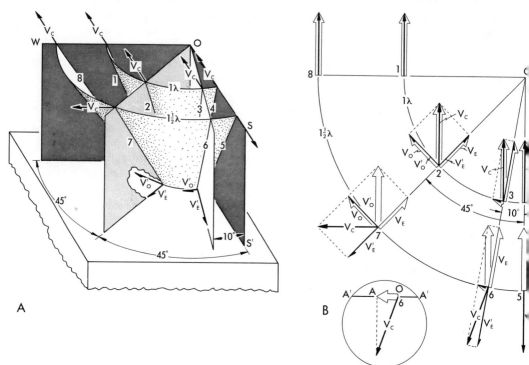

Fig. 7-16. (A) Enlarged view of a portion of the cones of equal retardation above the crystal in Fig. 7-13 B, assuming light of wavelength 550 mμ is the illuminant. The $1\frac{1}{2}\lambda$ cone and the extensions of four principal planes above the crystal have been added, plane OSS' being parallel to the privileged direction of the polarizer. The vibration directions of the E and O rays, V'_E and V'_O, are shown at their points of emergence at the crystal's upper surface. After emergence, each E- and O-ray pair traveling along ray paths 1 to 8 mutually interfere to produce a new vibration direction, V_C. (B) Full view of the horizontal plane WOS of Fig. 7-16A. Each ray's history is shown vectorially at its point of emergence in the plane. Thus the single-barbed hollow arrows indicate the vibration directions of the O ray and E ray—that is, V_O and V_E—into which the incident light from the polarizer (double-barbed hollow arrow) is resolved upon entering the crystal at its under surface. Single-barbed solid arrows V'_E and V'_O represent the vibrations of the E and O ray just as they emerge from the crystal's upper surface. The double-barbed solid arrow V_C indicates the vibration direction for the ray emerging from the crystal, this ray being the resultant of the interference of the E and O rays upon emergence. The circular inset indicates OA, the vibration direction and amplitude of the component of arrow V_C that is transmitted by the analyzer (privileged direction $A'A'$) when ray 6 passes into it.

Rays that travel parallel to principal planes that are at 0 or 90 degrees to the polarizer's privileged direction will vibrate parallel to this direction both during and after passage through the crystal. Rays 4 and 5 are examples; their parent rays from the polarizer formed only N–S vibrating E rays upon entry into the crystal. Rays 1 and 8 are also examples; they are the result of rays from the polarizer that formed only N–S vibrating O rays during passage. Thus the rays emerging along the cross hairs are entirely extinguished by the analyzer. For rays emerging along radii at small angles to the cross hairs, V_C deviates little from N–S; consequently, these rays are almost completely extinguished by the analyzer. Thus, for a uniaxial crystal whose optic axis emerges at O (Fig. 7–17), wedge-

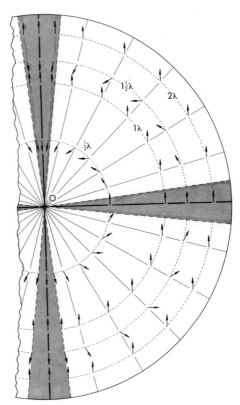

Fig. 7-17. Vibration directions of the resultant rays at their points of emergence in a uniaxial interference figure (slightly more than the right half of the field of view shown). The melatope is located at O, the cross-hair intersection. All rays emerging on or near the cross hairs vibrate in the same plane (N-S) as light from the polarizer. Such rays are extinguished by an E-W analyzer and their points of emergence are thus marked by dark areas—that is, isogyres—that are roughly wedge shaped as shown. The 1λ and 2λ circles, if monochromatic light were the illuminant, would be the sites of emergence of only N-S vibrating rays and would thus be marked by rings of blackness (not shown).

shaped, black areas (isogyres) mark the points of emergence of these rays. The curvature of the boundaries of these isogyres, which is particularly apparent toward the edge of the field of view, results from rotation of the vibration directions of the resultant rays caused by reflectional losses at lens surfaces. Kamb (1958) discusses the causes for this curvature in detail.

If monochromatic light is the illuminant and the path differences for this light are as marked in Fig. 7–17, not only the wedge-shaped areas but also the $n\lambda$ retardation cones will be areas of extinction. If white light is the illuminant, only the isogyres will persist as areas of blackness since the $n\lambda$

cone for one wavelength in the white light will often coincide with the $(n + \frac{1}{2})\lambda$ cone for another wavelength in this white light. Hence, with white light, no black circles indicating the emergence of $n\lambda$ cones in the field of view will be observed. Rays emerging intermediate between the n and $(n + \frac{1}{2})\lambda$ circles will be polarized either elliptically or circularly; these are not dealt with here.

The central line through an isogyre in an interference figure marks the trace of a principal plane that is either parallel or perpendicular to the vibration direction of light from the polarizer. The intersection between two central lines marks the point of emergence in the field of view of rays that, while in the crystal, traveled along its optic axis.

TYPES OF UNIAXIAL INTERFERENCE FIGURES

The appearance of a uniaxial interference figure depends upon whether the grain rests with its optic-axis direction (1) perpendicular, (2) at an oblique angle, or (3) parallel to the plane of the stage. The types of figures resultant from such orientations are respectively called (1) centered optic-axis figures, (2) off-centered optic-axis figures, and (3) the uniaxial flash figure. Their appearance and means of recognition will be discussed next.

Centered Optic-axis Figures

Centered optic-axis figures have been used exclusively in the previous discussions on the origin of the isogyres and isochromes in a uniaxial interference figure. Fig. 7–12, for example, was a typical, centered optic-axis figure. The *sine qua non* of a centered optic-axis figure is the location of the melatope (which marks the outcrop of the optic axis) at the cross-hair intersection. A necessary corollary is that the N–S and E–W bars of the uniaxial cross be respectively bisected by the N–S and E–W cross hairs. Rotation of the microscope stage will produce no observable change in the appearance of a precisely centered optic-axis figure. Thus, no matter how the stage holding the crystal in Fig. 7–18 is rotated, one of the innumerable principal planes that radiate outward from the optic axis

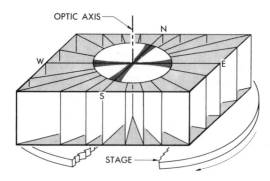

Fig. 7-18. Constancy of position of the isogyres (during rotation of the stage) in the interference figure of a crystal whose optic axis is normal to the microscope stage. A few of the innumerable principal planes hinging on the optic axis are shown.

(only a few of which are shown) will always intersect the field of view along each cross hair and be the site of an isogyre. Regardless of rotation of the stage, therefore, the isogyres will remain in the positions shown in Fig. 7–18.

Off-centered Optic-axis Figures

These figures are produced if the optic axis deviates from perpendicularity to the stage by the angle ν (nu) (Fig. 7–19A). The melatope, consequently, no longer coincides with the cross-hair intersection. Moreover, during a rotation of the stage, the melatope is observed to rotate in the same direction (Figs. 7–19B, C). The isogyres, meanwhile, retain (as always) an approximate parallelism to the cross hairs, their mutual intersection coinciding at all

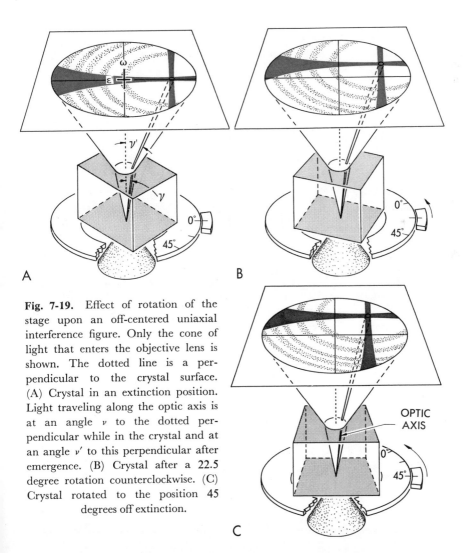

A

B

Fig. 7-19. Effect of rotation of the stage upon an off-centered uniaxial interference figure. Only the cone of light that enters the objective lens is shown. The dotted line is a perpendicular to the crystal surface. (A) Crystal in an extinction position. Light traveling along the optic axis is at an angle ν to the dotted perpendicular while in the crystal and at an angle ν' to this perpendicular after emergence. (B) Crystal after a 22.5 degree rotation counterclockwise. (C) Crystal rotated to the position 45 degrees off extinction.

C

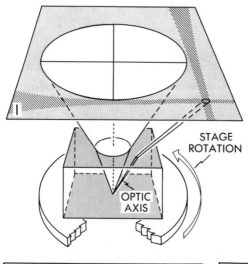

Fig. 7-20. (I) Crystal that produces an off-centered optic axis figure for which the melatope falls outside the field of view. (II), (III), and (IV) Motion of isogyre as the crystal is rotated counterclockwise. The opposite ends of the visible isogyre—that is, H and A—move respectively from H_{II} to H_{IV} and from A_{II} to A_{IV} with rotation. Since H_{II} to H_{IV} involves a motion in the same direction as the crystal was rotated (counterclockwise in this example), the H end is called the homodrome end. The end labeled A is called the antidrome end of the isogyre since motion A_{II} to A_{IV} is opposite to the rotation direction of the crystal.

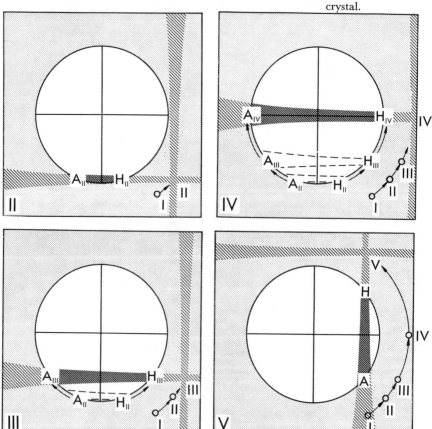

times with the melatope. The isochromes again appear as circles centered on the melatope. The isogyres and isochromes have the same origin as in centered optic-axis figures. The isogyres mark the traces within the field of view of principal planes that are parallel or nearly parallel to the nicol's privileged directions. Thus (as per the discussion on pp. 115-6) these traces are sites of complete extinction.

Note that if v' represents the angle between the perpendicular to the stage and the path in air of light that in the crystal had traveled along the optic axis (Fig. 7-19A), then from Snell's law

$$\omega \sin v = \sin v'$$

Often, however, the optic axis is so tilted that it falls outside the cone of light that enters the objective from the crystal. For this case, the melatope falls outside the field of view of the interference figure (Fig. 7-20 I). Even if no isochromes are present, the quadrant of the field of view that would contain the melatope, if the field were sufficiently enlarged, can readily be determined. Rotate the crystal of Fig. 7-20 I counterclockwise and observe the interference figure. An isogyre enters the field of view as in Fig. 7-20 II. With continued counterclockwise rotation of the crystal, the interference figure successively takes on the appearance of Fig. 7-20 III, IV, and V. Of the isogyre segments in the field of view, label the end nearest the melatope with the letter H and the end farthest from the melatope with the letter A. During the counterclockwise rotation of the crystal, note how H, one of the isogyre's intersections with the edge of the field of view, also moves counterclockwise (from H_{II} to H_{III} to H_{IV} in Fig. 7-20 IV) whereas the isogyre's opposite end moves clockwise (from A_{II} to A_{III} to A_{IV}). The H end, since it always moves in the same direction as the direction of turn of the crystal, is called the *homodrome end* (Gr., same course or path) of the isogyre; the A end, since it moves opposite to the crystal rotation, is called the *antidrome end* (Gr., opposite path).

A melatope, outside the field of view, is always located in that quadrant which contains the homodrome end of any visible portion of an isogyre. If segments of circular isochromes are present in the field, the melatope will be located in the quadrant containing the center for these circles.

Flash Figures

A flash figure is the interference figure produced from a crystal whose optic axis is parallel (or nearly parallel) to the plane of the microscope stage (Fig. 7-21A). The rays entering the crystal from the illuminating cone travel within an infinite number of principal planes that fan outward from the optic axis; only a few of these planes are shown in the figure. These planes emerge within the plane of the interference figure as a series of parallel traces (dashed lines in Fig. 7-21B). The O rays vibrate perpendicular and the E rays essentially parallel to these traces. The vibration directions for all rays emerging in the field of view are therefore essentially parallel to each other. Consequently, if the crystal is rotated until its optic axis is parallel to a nicol privileged direction, the field of view is almost completely filled with a diffuse, black cross (Fig. 7–21C). Only at the

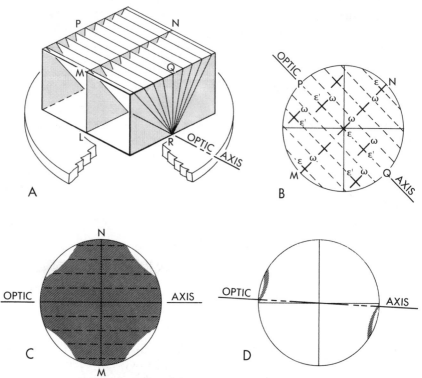

Fig. 7-21. Origin of the uniaxial flash figure. (A) Crystal with optic axis ideally oriented to yield a uniaxial flash figure. The planes hinging at the optic axis are principal planes. (B) Mutual parallelism of the traces of these principal planes within the field of view of the interference figure, the optic axis being at 45 degrees off extinction. The vibration directions of a few rays are shown where they emerge within the field. Note the parallelism of these vibrations. (C) Appearance of the isogyre at extinction position. (D) Appearance of the isogyre if the crystal is rotated slightly off extinction.

extreme NE, SE, SW, and NW edges of the field of view is nonextinction observed, an indication that the vibrations for rays emerging within these regions were not parallel to the vibrations of the majority of rays in the field. Upon turning the crystal only a degree or so, the diffuse cross is rapidly resolved into hyperbolic isogyres that are said to leave the view within the quadrants containing the optic axis after this rotation (Fig. 7–21D). The direction of movement of these isogyres, however, is sometimes difficult to determine.

The distribution of the isochromes in a uniaxial flash figure at 45 degrees off extinction is shown for a crystal of relatively high retardation (Fig. 7–22A)

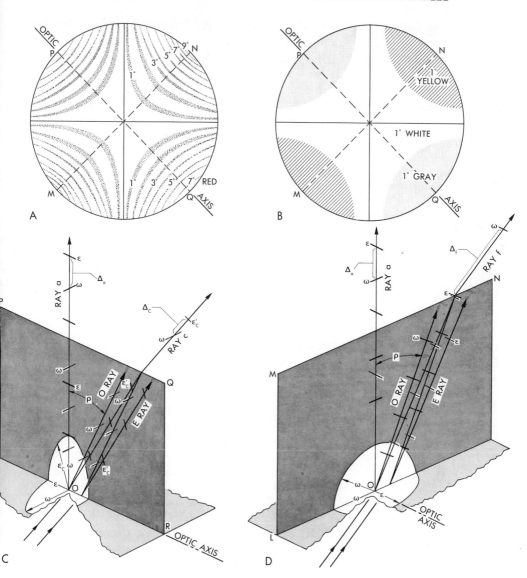

Fig. 7-22. (A) Distribution of the isochromes for a uniaxial flash figure at 45 degrees off extinction for a mineral of high birefringence. (B) Same for one of low birefringence. Note that outward from the cross-hair intersection the interference colors become progressively higher toward points M and N whereas toward points P and Q they become lower order; these areas of lower order colors mark the quadrants in which the optic axis lies. The reason for a lower retardation color at Q than at N is illustrated in (C) and (D), where planes PQR and LMN represent cross sections through Fig. 7-21A. In both (C) and (D) the refractive index value and vibration direction for each ray is shown by vectors that are perpendicular to or lie within the dark-shaded vertical planes. The intersections of these vertical planes with the crystal's indicatrix are unshaded.

and for one of low retardation (Fig. 7–22B). The high retardation of the first crystal may be due to its being excessively thick or its having a very high birefringence ($\epsilon - \omega$). Going outward from the cross-hair intersection in a flash figure at 45 degrees off extinction, the interference colors become increasingly higher in order toward the edges of two quadrants—for example, toward M and N in Fig. 7–22B—but lower in order toward the edges of the quadrants in which the optic axis lies—that is, toward P and Q in Fig. 7–22B. For very thick sections or minerals of high birefringence such as would yield figures like that in Fig. 7–22A, the decrease in order toward P and Q may be arrested and, as the edge of the field is approached, give place to a rise in order. This fall in retardation colors outward from the center, whether continuous or temporary, serves to identify the quadrants containing the optic axis.

To understand the phenomena just described, consider the interference between the E and O rays that emerge from a conoscopically illuminated crystal whose optic axis is parallel to the microscope stage. Within a vertical plane parallel to the optic axis (PQR in Fig. 7–22C), note that the vertical ray a is the result of interference between an O and an E ray that traveled perpendicular to the optic axis. The retardation for this ray is thus $\Delta_a = (\epsilon - \omega)t$, where t is the crystal thickness. Rays in plane PQR that emerge nearer the edge of the field of view in the interference figure—ray c for example—result from the interference between an E and an O ray that traveled more nearly parallel to the optic axis than did ray a. The retardation between these interfering rays is less than Δ_a because, although they travel a longer distance in the crystal, the difference in their indices, $(\epsilon'_c - \omega)$, is a much smaller value than $(\epsilon - \omega)$. Thus PQ, the trace of plane PQR in the interference figure, is the site of interference colors that decrease in order outward from the center. Sometimes, particularly in thick or highly birefringent crystals, the increase in path length for rays at the larger inclinations from the vertical more than offsets the decrease in birefringence, and as a result, the retardation colors begin to increase in order.

For a plane like LMN (Fig. 7–22D), which is perpendicular to the optic axis, all E rays traveling within it vibrate parallel to the optic axis and are therefore associated with a refractive index ϵ. Consequently, the birefringence between all E and O ray pairs that interfere after their emergence from plane LMN of the crystal remains as ($\epsilon - \omega$), regardless of their inclination from the vertical. With increased inclination, the path distance of these rays within the crystal, and consequently their retardation, increases. For ray f in Fig. 7–22D, for example, the retardation is approximately $\Delta_f = (\epsilon - \omega)(t/\cos\rho)$, where ρ may be taken to be the angle of inclination from the vertical of either the E or O ray that interfere to form ray f; the angle between these two rays is usually quite small. Thus Δ_f increasingly exceeds Δ_a in value as the angle ρ increases. Consequently, MN, the trace of plane LMN in the flash interference figures, is the site of interference colors that increase in order outward from the center.

EIGHT — Optical Examination
OF UNIAXIAL CRYSTALS

PREPARATION OF THE SAMPLE

The crushed crystals should be sieved to isolate the 100 to 120 mesh fraction for study. Grain thicknesses, as a result, should most frequently fall within the 0.125 to 0.149 mm range. Oil mounts of these crushed grains are prepared as previously described (p. 60).

ESTABLISHMENT OF UNIAXIALITY

The mineral grains in the oil rest on various fracture surfaces which, unless there is a dominant cleavage, will be at all manner of angles to the optic axis. Those grains for which this surface of rest is most nearly perpendicular to their optic axis will most likely yield an interference figure whose melatope is within the field of view. With the melatope thus visualized, the uniaxial nature of the crystal yielding the interference figure can readily be established since the Formée cross (formed by the isogyre intersection) will remain intact—that is, will not break up into a hyperbola—during rotation of the stage.

The grain most likely to yield such a near-centered optic axis figure is determined by examining the powder mount between crossed nicols with a low-power objective. Such a grain is characterized by (1) being one of the larger grains, yet (2) possessing a lower interference color at its center or thickest part than the other grains. For such grains, the ϵ' index will necessarily be close in value to ω. If these values are very close to each other, the grain's center will show a first-order black interference color in all positions of rotation (since $(\epsilon' - \omega)$ will be minimal). Such a grain most likely rests on a surface nearly parallel to the circular section. Consequently, the melatope of its interference figure will very likely be centered (or nearly centered) in the field of view.

COMPENSATION

Accessory Plates and Wedges

Three accessories are commonly used in optical mineralogy: (1) the simple quartz wedge (or a variant), (2) a first-order red or gypsum plate, and (3) a quarter-wave mica plate. Each is designed to slip into the accessory slot of the microscope so as to intercept and transmit all light rays. Like all anisotropic plates, each possesses two, mutually perpendicular, privileged directions. The one corresponding to the larger index N is called the slow direction; the one corresponding to the smaller index n, the fast direction. Generally, only the N privileged direction is indicated on the accessory's metal mount (by an etched line or arrow). Light waves that vibrate parallel to this N direction while passing through the accessory plate travel more

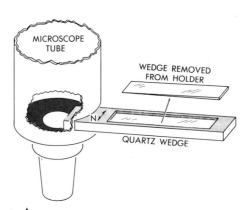

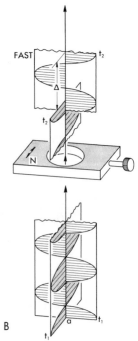

A

Fig. 8-1. (A) Quartz wedge being inserted, thin edge first, into the compensator slot of the microscope. Toward its thick edge the retardation produced by the quartz wedge increases to about 1700 mμ for some makes or to about 2800 mμ for others. (B) Compensator plate intercepting two perpendicularly polarized light waves that are here assumed not to interfere. After passage through the plate, the wave that vibrated parallel to the N direction of the plate (double headed arrow) is, at the time t_2, a distance $\varDelta$ behind the other. If the compensator plate is a first-order red (gypsum) plate, $\varDelta$ equals approximately 550 mμ. If it is a one-quarter wave (mica) plate, $\varDelta$ equals approximately 150 mμ. The double-headed arrow indicating the slow or N direction is sometimes labeled ζ or γ by the manufacturers.

slowly than do those that vibrate parallel to the unmarked n direction during their transmission. Consequently, after emergence from the accessory, the slow wave is retarded with respect to the fast wave (see Fig. 7–3).

A quartz wedge, if gradually inserted, thin end first, into the accessory slot (Fig. 8–1A), produces increasingly higher retardations as its thicker portions successively move into the light path. The first-order red or gypsum plate, on the other hand, is of constant thickness throughout. Its birefringence $(N - n)$ and thickness are such that it produces a retardation $(\varDelta)$ of 550 mμ. Consider, for example, two mutually perpendicular waves, precisely in phase at time t_1 in Fig. 8–1B (their mutual interaction being here ignored). After passing through the plate, they are no longer in phase. The vibration at time t_2 for the wave that traveled more slowly through the plate occurs precisely 550 mμ behind the vibration of the faster wave at this time.

The quarter-wave mica plate is generally a thin plate of mica of sufficient thickness and birefringence $(N - n)$ to produce a retardation $(\varDelta)$ of about 150 mμ (roughly one quarter of the wavelength of sodium light). Thus, if two mutually perpendicular waves are exactly in phase before passage through the plate, the wave that travels more slowly through the mica plate is retarded approximately 150 mμ behind the fast one after their emergence.

Addition

Suppose that an anisotropic crystal viewed at 45 degrees off extinction between crossed nicols exhibits a first-order gray retardation color $(\varDelta = 100$ m$\mu)$. Insertion of the gypsum plate so that its N direction is parallel to that of the crystal (Fig. 8–2A) changes the observed retardation color to second-order blue $(\varDelta = 650$ m$\mu)$. When insertion of an accessory increases the order of an interference color by an increment comparable to the retardation value of the accessory, the process is called *addition*. The explanation is as follows (Fig. 8–2A): After emergence from the crystal, the wave that had vibrated parallel to N_c in the crystal is 100 mμ behind that which had vibrated parallel to n_c. Upon entering the gypsum plate, this already retarded wave, since it vibrates parallel to N while within the gypsum plate, is retarded an additional 550 mμ, finally lagging 650 mμ behind the fast ray. If viewed between crossed nicols, therefore, an interference color corresponding to 650 mμ (that is, second-order blue) is observed.

Subtraction

Assume the crystal to have been rotated 90 degrees so that its n and N privileged directions have exchanged positions (Fig. 8–2B). The retarded wave that vibrated within the crystal parallel to N_c, the crystal's slow direction, would be as before 100 mμ behind the fast wave upon emergence from the crystal. Now, however, when it enters the gypsum plate, this formerly slow wave vibrates parallel to the fast direction whereas the formerly fast wave vibrates parallel to the slow direction of the gypsum plate. Consequently, although the coarsely ruled wave (Fig. 8–2B) was 100 mμ ahead prior to entering the gypsum plate, the finely ruled wave gained 550 mμ to overtake and surpass it while traveling through the

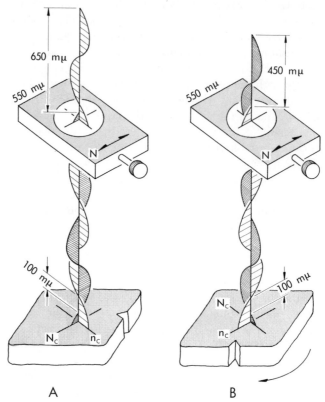

A   B

Fig. 8-2. Crystal of retardation 100 mμ at 45° off extinction in additive position (A) and subtractive position (B) with respect to the first-order red (550 mμ) plate. Note that the finely ruled wave, which traveled slower in the crystal, is further retarded by the gypsum plate in (A). In (B) the wave that traveled faster (coarsely ruled) is retarded by the gypsum plate so that the formerly slow wave has overtaken and passed it by 450 mμ.

gypsum plate. Thus, on emergence from the gypsum plate, the finely ruled wave is ahead by 450 mμ (550 mμ − 100 mμ). The retardation color corresponding to 450 mμ—that is, 1° orange—is thus seen. Such a process, produced because the fast wave in the crystal becomes the slow wave in the gypsum plate, is called *subtraction;* the resultant retardation equals the difference between the individual retardation values of the two plates involved.

General Rules

Addition occurs when the N directions of the crystal and of the compensator plate most nearly coincide. Subtraction occurs when the N directions of the crystal and of the compensator are at right angles. Stated conversely, if addition occurs, then N of the crystal and of the compensator must be parallel or nearly parallel. If subtraction occurs, they must be perpendicular to each other or nearly so.

If, after insertion of a 1° red plate or quarter-wave plate—or during insertion of a quartz wedge (thin end first)—interference colors of higher order are observed to displace those of lower order, addition has occurred. If lower order colors displace those of higher order, subtraction has occurred.

DETERMINATION OF OPTIC SIGN

Principles Involved

The orthographically projected vibration directions of the E and O rays emerging at common points in the field of view are drawn in Fig. 8–3 for both a uniaxial $(+)$ and a uniaxial $(-)$ interference figure. By definition, ϵ' exceeds ω in $(+)$ crystals but is less than ω in $(-)$ crystals. Hence in Fig. 8–3, where the relative values of ϵ' and ω for these rays are schematically illustrated by the lengths of lines indicating their vibration directions, the radial E-ray vibration—that is, the vibration that "points" toward the melatope—is longer than the O-ray vibration for the $(+)$ crystal but

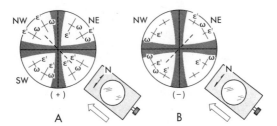

Fig. 8-3. Angular relationship of N direction of compensator plate to the N and n vibrations (respectively shown as long and short lines) for rays emerging within the four quandrants into which the uniaxial cross divides the interference figure. These quadrants are labeled with the compass directions usually used to refer to them. In (A) the crystal is positive $(N = \epsilon', n = \omega)$; in (B) it is negative $(N = \omega, n = \epsilon')$. For illustrative purposes, the two quadrants in which subtraction occurs after insertion of the accessory are joined by a dashed line.

shorter for the (—) crystal. Consequently, insertion of an accessory with its
N direction as marked in Fig. 8–3 produces subtraction in the quadrants
to the NW and SE of the melatope for (+) crystals or to the NE and SW
for (—) crystals. In the remaining quadrants addition occurs. Useful rules
for determining the optic sign are: (1) If the line joining the quadrants of
subtraction (dashed in Fig. 8–3) is perpendicular to the N direction of the
accessory plate—thus forming an imaginary plus sign—the crystal is (+)
in optic sign. (2) If this line is parallel—forming an imaginary negative
sign—the crystal is (—) in optic sign.

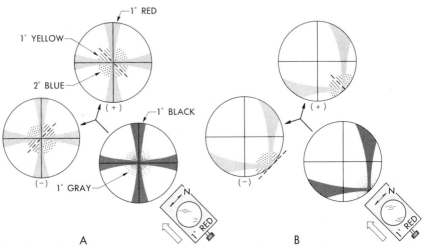

Fig. 8-4. Determination of optic sign from the disposition of interference colors
after insertion of the 1° red (gypsum) plate. Note that results are similar whether the
optic axis is precisely centered (A) or not (B). The heavy dashed line joining the
quadrants in which subtraction occurs—that is, those containing 1° yellow—has
been added for illustrative purposes.

The optic sign can thus be determined by locating the quadrants in
which subtraction occurs upon insertion of an accessory whose N direction
is known. For interference figures in which few isochromes appear, the
first-order red (gypsum) plate or the quarter-wave mica plate serve equally
well for determination of the optic sign. If the first-order red plate ($\Delta = 550$ mμ) is inserted, the 1° gray ($\Delta = 125$ mμ) areas adjacent to the melatope
become 1° yellow ($\Delta = 425$ mμ; that is, $125 - 550$ mμ) in the quadrants
of subtraction but 2° blue ($\Delta = 675$ mμ; that is, $125 + 550$ mμ) in the
regions of addition. If the interference figure is viewed after insertion of the
quarter-wave mica plate ($\Delta = 150$ mμ), however, these same 1° gray areas
become 1° black ($\Delta = 25$ mμ; that is, $125 - 150$ mμ) in the quadrants of
subtraction but 1° white ($\Delta = 275$ mμ; that is, $125 + 150$ mμ) in the
quadrants of addition. Consequently, 1° yellow on opposite sides of the
melatope (Fig. 8–4) marks the quadrants of subtraction if a first-order red
plate is inserted. If a quarter-wave mica plate is inserted, two black dots

(Fig. 8–5) mark the quadrants of subtraction. Thus, following the rules of the preceding paragraph, if the line joining the centers of these two yellow areas or black dots is perpendicular to the N direction of the inserted accessory, the crystal is $(+)$ in optic sign. If parallel, the crystal is $(-)$.

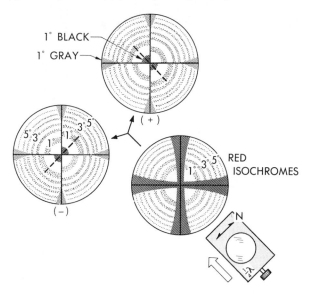

Fig. 8-5. Determination of optic sign from the disposition of the two black dots produced by insertion of the quarter-wave mica plate. The two black dots mark the quadrants in which subtraction occurs. A heavy dashed line joining the quadrants in which subtraction occurs has been added for illustrative purposes.

For interference figures containing numerous, closely spaced isochromes, the quartz wedge has certain advantages in optic sign determination, particularly for off-centered figures in which the melatope does not appear in the field of view. Observed while the quartz wedge is being inserted thin edge first, the isochromes move outward from the melatope in the quadrants of subtraction (Fig. 8–6) and inward toward the melatope in the quadrants of addition. If the line joining the quadrants of subtraction on opposite sides of the melatope is perpendicular to the N direction of the wedge, the crystal is $(+)$ in sign (Fig. 8–6A); otherwise (Fig. 8–6B), it is $(-)$.

To determine the optic sign in off-centered interference figures, one needs to determine whether ϵ' exceeds ω or not. To do this, first rotate the stage until a line connecting the melatope to the cross-hair intersection would be approximately parallel to the N direction of any inserted accessory. The crystal is now oriented with its ϵ' privileged direction parallel to (and its ω privileged direction perpendicular to) the N direction of the accessory (Fig. 8–6C, D). Insertion of the quartz wedge over the crystal in this position thus produces addition for $(+)$ crystals (since for them ϵ' exceeds ω) and subtraction for (−) crystals (since for them ϵ' is less than ω). As Fig. 8–6C illustrates, addition in this interference figure is expressed by the higher order isochromes displacing the lower order ones—that is, an inward movement toward the melatope. In Fig. 8–6D subtraction is expressed by the lower

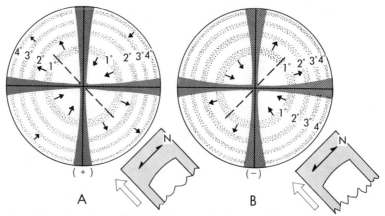

Fig. 8-6. Movement of the isochromes if viewed while a quartz wedge is inserted, thin end first. (A) Centered, positive, uniaxial figure. (B) Centered, negative figure. (C) Off-centered, positive figure. (D) Off-centered, negative figure. For illustrative purposes the quadrants of subtraction in (A) and (B) are joined by dashed lines.

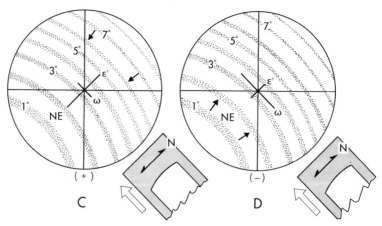

order isochromes displacing the higher order ones, that is, a movement outward from the melatope. If the movement of the isochromes in the interference figure is hard to observe, it is sometimes preferable to convert to the orthoscopic set-up and observe whether insertion of the quartz wedge causes addition or subtraction with respect to the grain image. Addition is expressed by movement of the isochromes outward from the grain's center, subtraction by their movement inward towards the grain's center (see Fig. 8–9).

MEASUREMENT OF REFRACTIVE INDICES

The value of the refractive index ϵ' of a uniaxial grain depends upon the angle between its optic axis and the plane of the glass slide on which it rests. Fig. 8–7 illustrates the variation of ϵ' and retardation color for

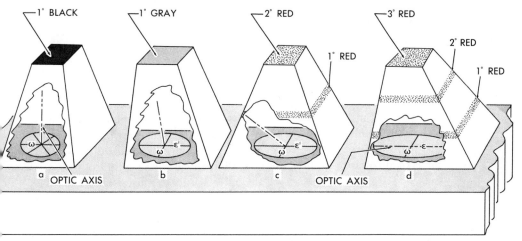

Fig. 8-7. Relationship between the orientation of the optic axis (dash-dotted line) and the observed interference color for four grains of equal thickness. Note that, as the optic axis becomes increasingly parallel to the glass slide (from grains a to d) ϵ' approaches true ϵ in value and the retardation attains its highest order for that thickness. Only the red isochromes (stippled) and first-order black and gray are shown.

four grains of approximately equal thickness of a given uniaxial $(+)$ crystal, the privileged directions being drawn as lengths proportional to the values of ϵ' and ω for that grain. For grain b, whose optic axis is nearly perpendicular to the slide, ϵ' is close to ω in value—a fact testified to by its relatively low retardation color. The optic axis in grains c and d is increasingly parallel to the slide; consequently, ϵ' becomes increasingly closer to ϵ in value and, as a result, the retardation colors become higher in going from grain a to d. For grain d, the optic axis is parallel to the slide; as a consequence, ϵ' now equals ϵ and grain d therefore possesses the highest interference color of all the grains (since its birefringence for that position equals $(\epsilon - \omega)$, the maximum possible).

In a given powder mount, there is usually a multiplicity of grains such as b and c; measurement of ϵ' for them is of no particular value. An exception occurs in cases like the rhombohedral carbonates, where a strong rhombohedral cleavage (1011) closely controls the plane upon which the grain will rest. Since, from one mount to the other, grains will rest on this particular cleavage plane, the ϵ' for such grains (symbolized $\epsilon'_{10\bar{1}1}$) may be measured and will yield information as to which rhombohedral carbonate is involved.

The indices ω and ϵ, in contrast to ϵ', are always definitive values whose determination will aid in identifying the mineral. Their determination, therefore, will be considered in detail.

Measurement of ω

One of the two privileged directions of a uniaxial grain always corresponds to ω (for example, in grains a-d in Fig. 8–7). Thus, any grain can be selected for the measurement of ω; however, one of relatively low interference color (for example, grain a or b) will yield a more readily interpreted interference figure. Rotate the microscope stage until an isogyre of this interference figure is bisected by the E–W cross hair as in Fig. 8–8A or B. For illustrative purposes, in each figure the privileged directions that are associated with the central ray of the cone of illumination are shown at this ray's point of emergence at the cross-hair intersection. The path of this central ray and therefore its associated privileged directions within the crystal coincide in direction with those of all the rays that pass through the crystal after conversion of the microscope to the orthoscopic set-up

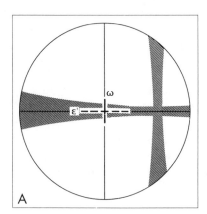 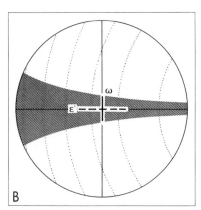

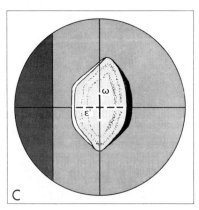

Fig. 8-8. (A and B) Correct conoscopic orientations of grains so that their ω privileged direction is N-S. (C) Use of oblique illumination to test the value of ω against the index of the immersion oil for the grain that was conoscopically oriented in (A) or (B). Assuming a N-S polarizer, note that only O rays, for which the crystal's index is ω, are transmitted by the crystal if it is illuminated orthoscopically as in (C). Since no E rays are transmitted in this orientation, the ϵ' privileged direction is drawn as a dashed line.

(see Fig. 7–10). Thus, the center of the interference figure discloses both the direction and identity (that is, whether ϵ' or ω) of the privileged directions possessed by the grains if viewed orthoscopically (Fig. 8–8C).

Uniaxial grains oriented as in Fig. 8–8C transmit only O rays (if the polarizer's vibration direction is N–S). The grain's ϵ' privileged direction has therefore been drawn as a dashed line, because no light vibrating parallel to this direction is transmitted by the grain when in this orientation. Thus, ω is the only index exhibited by the grain in this orientation. Its value can be measured, after conversion to the orthoscope and removal of the analyzer, by either oblique illumination or the Becke line method in the same manner that n was for isotropic grains.

Measurement of ϵ'

Occasionally, as previously explained, it may be necessary to measure ϵ'. The grains of Fig. 8–8 would be positioned for this measurement if the stage were rotated until the N–S isogyre was bisected by the N–S cross hair. Conversion to the orthoscope and removal of the analyzer would then permit measurement of ϵ' by oblique illumination or Becke line observations.

Measurement of ϵ

Unlike ω, ϵ cannot be measured from every grain. Only a grain whose optic axis is parallel to the microscope stage (for example, grain d in Fig. 8–7) possesses an extraordinary privileged direction corresponding in index to ϵ. Grains of this orientation are located by scanning the slide (orthoscope; crossed nicols; low-power objective) to find the grain possessing the highest retardation color. The high retardation color is an indication that the difference between the refractive indices for the grain's two privileged directions is equal (or very close) to the maximum possible, $(\epsilon - \omega)$; in other words, its optic axis is parallel or nearly parallel to the microscope stage. An interference figure serves to check this orientation; a flash figure should be obtained if the grain is correctly oriented for measurement of ϵ.

The proper grain now selected, its ϵ privileged direction must be distinguished from its ω direction. This is done by viewing the grain orthoscopically between crossed nicols at 45 degrees off extinction during insertion of a gypsum plate or quartz wedge (Fig. 8–9). Using a quartz wedge two possibilities exist: (1) The high-order isochromes are observed to displace the low-order isochromes (addition), indicating that the N directions of accessory and grain are parallel (Fig. 8–9A); or (2) The low-order isochromes displace the higher (subtraction), indicating that the N direction of the accessory is parallel to the n direction of the grain (Fig. 8–9B). In either case, the observer, since he knows the N direction of the accessory, discovers the N direction of the crystal as well. However, by definition, $\omega > \epsilon' > \epsilon$ in uniaxial $(-)$ crystals whereas $\epsilon > \epsilon' > \omega$ in uniaxial $(+)$ crystals. Thus, if the observer knows the optic sign of the grain under observation, he knows whether the ϵ direction coincides with the N direction (as in a uniaxial positive crystal) or with the n direction (as in a uniaxial negative grain). The ϵ direction thus identified, the microscope stage is now rotated until the ϵ direction is N–S and at extinction, and the accessory is completely withdrawn. The upper nicol is then also withdrawn and the grain is in position for measurement of ϵ by the usual methods.

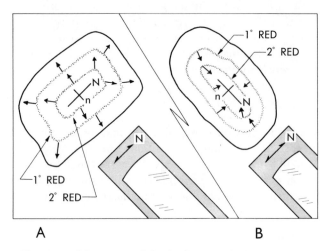

A   B

Fig. 8-9. Movement of the isochromes (only the red ones are shown) as the quartz wedge is inserted, thin edge first. (A) For a grain whose N direction parallels that of the wedge. Thus addition occurs and the higher order isochromes displace the lower order ones. (B) For a grain whose N direction is transverse to that of the wedge. For this grain the lower order isochromes displace the higher order ones—that is, subtraction occurs. If a gypsum plate were inserted instead of the quartz wedge, in (A) the 1° and 2° red isochromes would become 2° and 3° red, respectively, whereas in (B) they would become 1° black and 1° red, respectively.

If the optic sign of the grain has not been previously determined, the ϵ vibration direction can sometimes be determined from the flash figure at 45 degrees off extinction. As previously discussed (see p. 122), the interference colors in flash figures decrease outward from the center toward the quadrants bisected by the optic axis and, therefore, by the ϵ vibration direction. Once located, the ϵ vibration direction can be rotated N–S to permit the index ϵ to be measured from the grain.

Indirect Measurement of ϵ

Where a dominant cleavage greatly diminishes the possibility of fractures occurring parallel to the crystal's optic axis, direct measurement of ϵ becomes extremely difficult. However, it is possible to calculate ϵ from a measured value of ϵ' provided that (1) the value of ω is known and (2) the angle (within the crystal) between the optic axis and the normal to the stage can be measured from the interference figure. The value of ω is of course readily determined by immersion methods and, fortunately, the angle in question, labeled ν in Fig. 8–10, can be quickly measured by adapting a method described by Tobi (1956, p. 516).

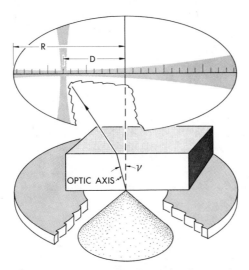

Fig. 8-10. A crystal whose size is greatly exaggerated is shown on a microscope stage, its front half removed for illustrative purposes. The angle between its optic axis and the perpendicular to the stage (dashed line) is labeled ν. The interference figure of this grain, if viewed with a micrometer ocular, permits measurement (in the scale units of this ocular) of D, the melatope's distance from the cross-hair intersection, and of R, the radius of the field of view.

To determine ν from the interference figure of the grain for which ϵ' has been determined, measure R and D (see Fig. 8–10) in terms of the units of a micrometer ocular. The values of D/R and of ω for the crystal are coordinates of a point on a diagonal line in Fig. 8–11 that indicates the value of ν.*

Knowing ω, ϵ', and ν for a particular grain, the value ϵ can be calculated as follows

$$\epsilon^2 = \frac{\epsilon'^2(\omega^2 \sin^2 \nu)}{\omega^2 - (\epsilon'^2 \cos^2 \nu)} \qquad \text{(Eq. 8–1)}$$

* The objective used to obtain the interference figure must possess a numerical aperture of 0.85 in order for Fig. 8–11 to yield correct values. For a different numerical aperture (A′), multiply observed D/R ratios by A′/0.85 to correct them for use in Fig. 8–11.

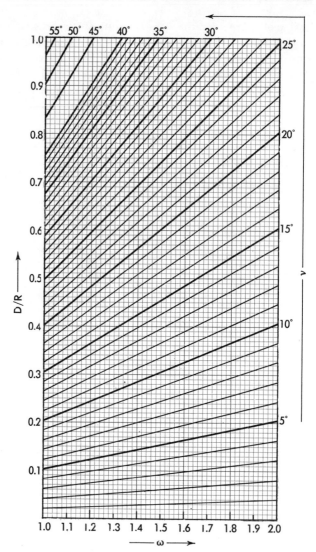

Fig. 8-11. Chart for the determination of ν, the angle between the optic axis and the normal to the stage, provided ω and D/R are known. (Adapted from Tobi, 1956.)

The calculations are tedious but fortunately a rapid and accurate solution is obtainable from an adaptation of a chart devised by Mertie (Fig. 8–12). An example best explains its use. For a given $(+)$ mineral, suppose $\omega = 1.550$, $\epsilon' = 1.600$, and, for the grain from which ϵ' was determined, $\nu = 35°$. On Fig. 8–12 locate ω on the left-hand scale and the point whose coordinates are ϵ' and ν in the scale's interior; note that the lower scale is used for ν. A line through these two points intersects the ϵ scale at the value of ϵ for the mineral in question—that is, about 1.720 for this example. Mathematically calculated from Eq. 8–1 the value is 1.719.

Fig. 8–12 can be used to determine ϵ, even if ϵ' has been measured from a grain whose melatope falls outside the field of view. In this case, the value of ν may be determined as follows: (1) Rotate the microscope

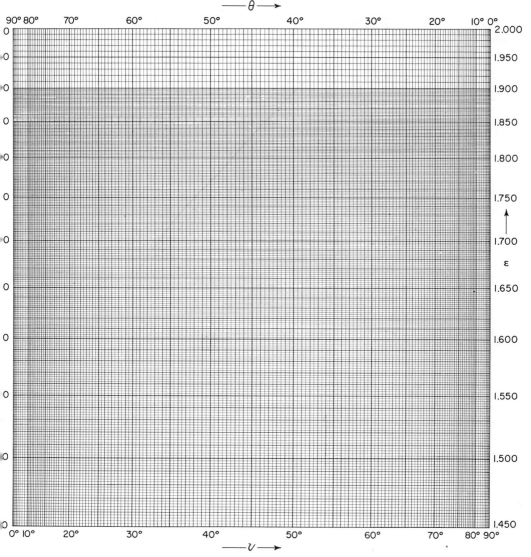

Fig. 8-12. Chart for rapid solution of Eq. 8-1 or Eqs. 6-3 and 6-4. Adapted from the chart for determining $2V$ by Mertie (1942, p. 542).

stage until the two opposite edges of one arm of an isogyre each successively touch the end of a cross hair (Fig. 8–13A and B). Average these two angular positions of the stage and record this value as S_1. (2) Repeat this process for the other arm of the same isogyre, the same end of the same cross hair again serving as the reference point (Fig. 8–13C and D). Record the average of these two stage positions as S_2. (3) Let ψ (psi) equal the smallest angle between stage positions S_1 and S_2. (4) The value of ν may now be determined from the point representing the grain's ψ and ω values on Fig. 8–14A.

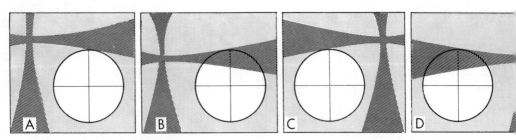

Fig. 8-13. Determination of the value of the angle ψ for an off-centered figure in which the melatope occurs outside the field of view. The stage is rotated until the interference figure resembles (A) and then (B). The average of the stage readings for these two positions is the stage position for which the central line through the E-W isogyre would be tangent to the north edge of the field of view. The average of the stage positions for (C) and (D) is the stage position for which the central line through the E-W isogyre is again tangent to the north edge of the field. The smallest difference between these two average positions is angle ψ.

Fig. 8–14A is applicable only to interference figures obtained with an objective whose numerical aperture is 0.85. Similar charts for objectives of differing numerical aperture may be constructed from the following equation

$$\tan \nu = \frac{\tan u'}{\cos \psi/2} \qquad \text{(Eq. 8–2)}$$

where u' represents arcsin N. A./ω.

It will be noted in Fig. 8–14A that no values of ψ greater than 140 degrees were calculated. This is principally because the accuracy of determining ν for these larger values of ψ drops off considerably. As an exercise, the reader is invited to test the accuracy of Fig. 8–14A by determining ν for a rhomb of calcite ($\omega = 1.658$) in an oil mount. If the grain is truly lying on a cleavage rhomb, not propped up on (or against) a neighboring grain, ν should be determined to be $44.5° \pm 3°$. Since the edge of some isogyres may be quite diffuse, care is needed in the measurement of ψ.

An example illustrates the convenience of the complete method. For a powdered unknown, ω is determined to be 1.679 and the optic sign $(-)$. In all oil mounts a dominant cleavage prevents grains from resting on other surfaces. In the absence of grains oriented with their optic axes parallel to the plane of the stage (from which ϵ could be measured directly), normal immersion method techniques are used, and the ϵ' index of several cleavage fragments is found to be 1.588. From the interference figure of this same grain, the method shown in Fig. 8–13 is employed four times

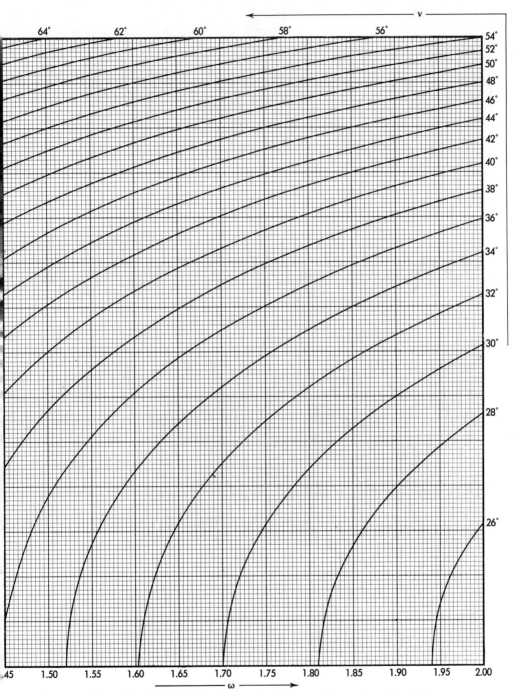

Fig. 8-14A. Chart for determining ν, the angle between the optic axis and the normal to the stage, provided ω and ψ are known.

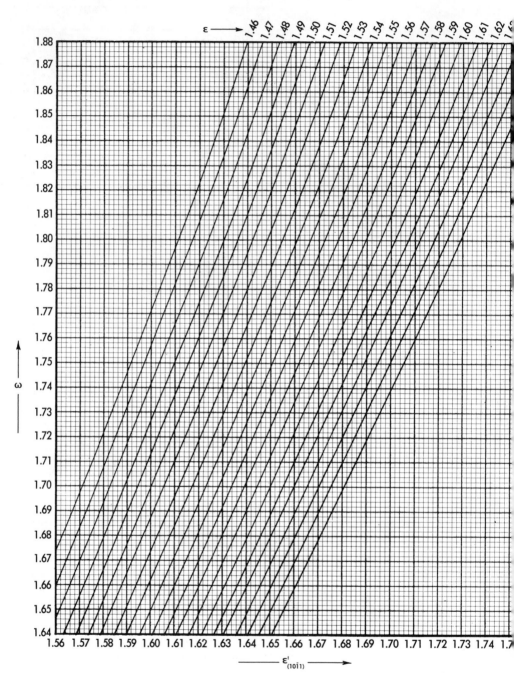

Fig. 8-14B. Chart for determining ϵ for the rhombohedral carbonates, provided ω and $\epsilon'_{(10\bar{1}1)}$ are known. (After Loupekine, 1947.)

(once for each end of the two cross hairs), care being taken not to jar the grain or the slide. The average of the four values of ψ thus obtained is 105°. From this value and that of ω, ν is determined to be 44° from Fig. 8–14A. Now, from these values [$\omega = 1.679$, $\epsilon' = 1.588$, $\nu = 44°$, optic sign $(-)$], ϵ is determined to be *ca.* 1.505 by means of Fig. 8–12. From these indices the mineral may be identified as dolomite ($\omega = 1.679$, $\epsilon = 1.502$).

The accuracy of determining ϵ in the foregoing method depends upon (1) the accuracy of measurement of ϵ', (2) the value of ν and the accuracy of its measurement, and (3) the value of ϵ/ω for the mineral. Assuming that ϵ' is measured to within 0.001 and that ν is about 45° (and is measured to ± 3°), ϵ can be determined to ± 1 percent if, for the mineral, ϵ/ω lies between 0.89 and 1.10 in value. As ϵ/ω approaches 1.00 in value, the error in determining ϵ decreases markedly to a fraction of 1 percent.

A method intended for the use of rhombohedral carbonates alone has been introduced by Loupekine (1947). In it, the measured values of ω and of $\epsilon'_{10\bar{1}1}$ (that is, ϵ' measured from a crystal resting on a 1011 cleavage) permit ϵ to be calculated to within ± 0.004 (Fig. 8–14B). Utilizing a correction factor (Loupekine, p. 504), this error is theoretically reduced to ± 0.001.

Dispersion

In advanced work the values of ϵ and ω for a mineral may be measured at different wavelengths of light. Such measurements on quartz (Winchell

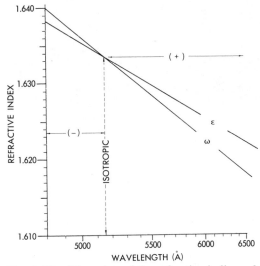

Fig. 8-15. Dispersion of the refractive indices of metatorbernite. Note that for one particular wavelength it is isotropic whereas it is optically positive for longer wavelengths and optically negative for shorter wavelengths.

and Winchell, 1951, p. 247), for example, indicate $\omega_F = 1.54968$, $\epsilon_F = 1.55898$; $\omega_D = 1.54425$, $\epsilon_D = 1.55336$; and $\omega_C = 1.54190$, $\epsilon_C = 1.55093$. Metatorbernite, $Cu(UO_2)_2P_2O_8 \cdot 8H_2O$, is unusual in this respect since it is uniaxial ($-$) for violet or blue light ($\omega_F = 1.638$, $\epsilon_F = 1.636$), isotropic for green light of wavelength 5120 Å ($\omega = \epsilon = 1.6335$), and uniaxial ($+$) for red light ($\omega_C = 1.618$, $\epsilon_C = 1.622$). Fig. 8–15 illustrates these relationships graphically for metatorbernite.

DETERMINATION OF RETARDATION AND BIREFRINGENCE

Estimation from Interference Colors

The absolute value of a mineral's birefringence, ($\epsilon - \omega$), can be quickly estimated from the highest order interference color observable among the smaller grains in a powder mount. Assuming that the grains in the mount represent the 100 to 120 mesh fraction of a thoroughly sieved sample, all have passed through openings of 0.149 mm but not through openings of 0.125 mm (Fig. 8–16). Their thickness (t) and width (W) are

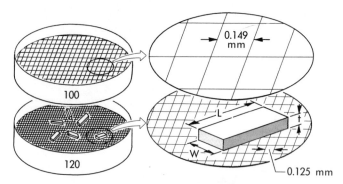

Fig. 8-16. *Left:* The 100 to 120 mesh fraction of a crushed mineral as caught on the 120 mesh sieve. *Right:* Enlarged view of a grain to show its length (L), width (W), and thickness (t) as it rests in its most stable position—that is, on its surface of maximum area. If sieving was vigorous and complete, its width must exceed 0.125 mm; otherwise it would have passed through the 120 mesh sieve. (Mesh openings not to scale.)

therefore undoubtedly less than the 0.149 mm openings. Assuming that the sieving was thorough,* the width of the grains as they lie on the glass slide

---

* If it was not, many very small grains capable of passing through the 0.125-mm openings will remain in the sieved sample.

most likely exceeds 0.125 mm; their thickness will, however, in many cases be less than 0.125 mm (particularly for flakes). However, for the narrowest grain showing the maximum interference color (of the hundreds on the slide) two assumptions appear likely : (1) The thickness, t, for this grain most closely approaches 0.125 mm (we shall assume it to be 0.12 mm*) and (2) its optic axis is parallel to the slide so that the indices for its privileged directions are ϵ and ω (or, if ϵ' and ω, then a value of ϵ' very close to ϵ). Using Fig. 8–17, $\varDelta$ can be evaluated to within 100 mμ from interpretation of the maximum interference color observed. Thus, with $\varDelta$ known and the thickness t approximately known from the minimum sieve opening, $(\epsilon - \omega)$ can readily be calculated from an adaptation of Eq. 7–7; that is,

$$(\epsilon - \omega) = \frac{\varDelta}{t} \qquad \text{(Eq. 8–3)}$$

Fig. 8–17 permits quick, graphic solutions of Eq. 8–3 as follows: (1) Locate on it the vertical color bar that corresponds to the maximum interference color observed in the grains. (2) Next locate the horizontal line corresponding to the crystal thickness, t. (3) The intersection of the vertical color bar and horizontal thickness line marks a point (or small area) in the interior of the color chart. (4) Through or near this point will pass several radial lines emanating from the origin of the color chart. (5) Follow these lines outward to the bold-faced numbers of the $(N - n)$ scale (the birefringence scale) to the top and right of the color chart. (6) The $(N - n)$ value thus indicated is likely to be quite close to $(\epsilon - \omega)$ for the mineral being studied. By way of example, assume that the 100 to 120 mesh fragments of a powdered unknown exhibit a maximum interference color of 2° green. Using Fig. 8–17, the value $(\epsilon - \omega)$ seems likely to fall within the 0.006 to 0.008 range.

For the sieved crystals of our example, the thickness of the crystal showing a maximum interference color was presumed to be near 0.12 mm. Fig. 8–17 is also especially useful for determining the birefringence $(N - n)$ of minerals in thin sections of rocks. Such wafers, cemented onto glass slides for study, are paper thin and generally 0.03 mm in thickness. Thus the intersection of the 0.03 mm thickness line with the particular vertical, color bar of Fig. 8–17 that corresponds to the maximum interference color observed for the mineral in the thin section would be used. Sometimes, however, thin sections vary in thickness from 0.03 mm, either throughout the section or in various areas; for example, the wafer may be slightly wedge shaped. In such cases the experienced petrographer can determine the thickness of particular areas of the thin section by observing the maximum interference color of any quartz grains $(\epsilon - \omega = 0.009)$ locally present. Thus the intersection of the vertical color bar (corresponding to this maximum interference color) with the radial line corresponding to an $(N - n)$ value

---

* If the grains of the mount represent the 100 to 200 mesh fraction (as some may prefer), this minimum thickness may be assumed to be about 0.07 mm.

of 0.009 falls on the horizontal line corresponding to the thickness of the thin section in Fig. 8–17.

The more important of the rock-forming minerals are arranged in Fig. 8–17 according to their maximum birefringence. Uniaxial minerals listed are denoted by (U), their birefringence corresponding to ($\epsilon - \omega$). For completeness, the biaxial minerals are also listed and denoted by (B). A (+) or (−) sign, if present behind U or B, indicates the mineral's optic sign. The birefringence ($N - n$) for biaxial crystals, as Chapter 9 will make clear, corresponds to ($\gamma - \alpha$). If it is included in the list of Fig. 8–17, a mineral can sometimes be rapidly identified on that basis alone. On the other hand, a particular value for maximum birefringence—for example, 0.010—may be possessed by several different minerals, some of which may not appear on the list. Consequently, if an initial identification from the list is made, it is wise to test its validity by further comparison of as many additional optical and physical properties as possible. Reference books such as Winchell and Winchell (1951) contain complete descriptions of minerals with which the properties of the unknown can be compared.

Precise Measurement

Retardations up to 2700 mμ can be quantitatively measured ($\pm$ 2 percent) by a Berek compensator or a slot compensator (Fig. 8–18).

Fig. 8-18. *Below:* Berek compensator (courtesy of E. Leitz, Inc.). *Right:* Slot compensator (courtesy of Bausch and Lomb Optical Company).

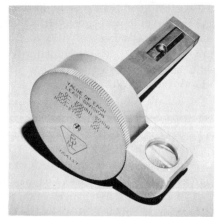

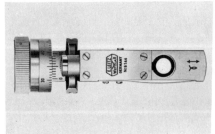

The latter is particularly convenient to use since readings are made directly in millimicrons, avoiding the necessity for conversion tables. The relatively inexpensive graduated quartz wedge manufactured by Cooke, Troughton, and Sims also permits fairly accurate measurement of retardation. Such precise measurements of retardation of a crystal plate, if followed by accurate measurement of its thickness, permit the birefringence to be accurately calculated (Eq. 8–3). Conversely, for plates of known birefringence an accurate measurement of their retardation permits their thickness to be calculated.

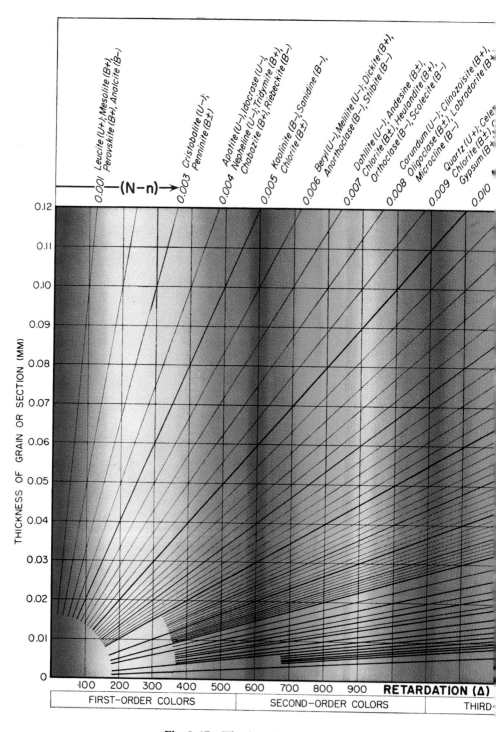

Fig. 8-17. The interference color associated with a particular value of retardation is shown as a vertical bar directly above that value. Thickness is plotted along the ordinate. The radial lines connect points of equal birefringence. As an example of the chart's use, note that the maximum

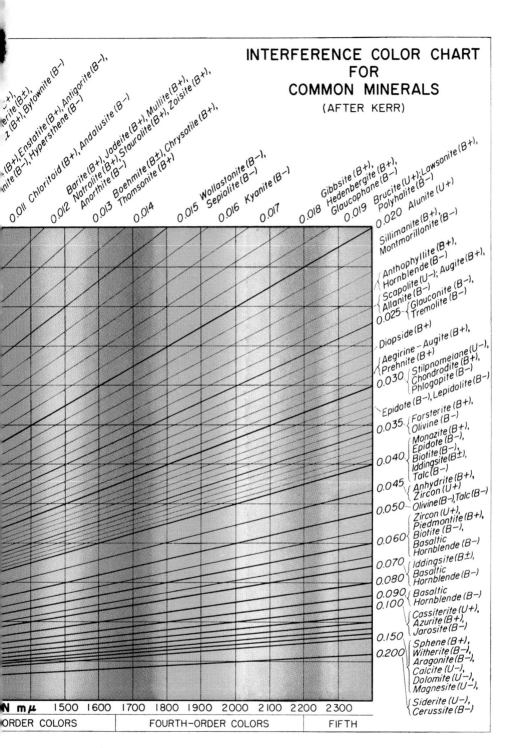

INTERFERENCE COLOR CHART
FOR
COMMON MINERALS
(AFTER KERR)

interference color for quartz (birefringence, 0.009) in a thin
section (thickness, 0.03 mm) is located at the intersection
between the 0.009 radial and the horizontal line corresponding to 0.03 mm. The color determined is just barely
first-order yellow.

EXTINCTION ANGLES: SIGN OF ELONGATION

The angle between a crystal plate's privileged directions and any linear crystallographic feature present on the plate is called its extinction angle. Such linear crystallographic directions may be the lines of intersection of a set of cleavage planes with the crystal plate—called cleavage traces (Fig. 8–19)—or it may be the straight boundary of the plate where it is

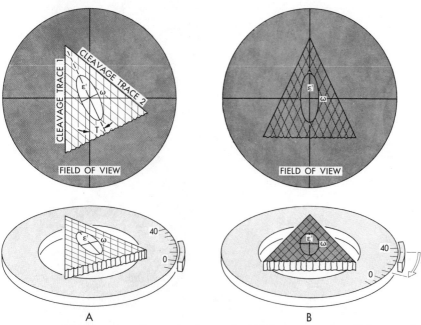

Fig. 8-19. Measurement of an extinction angle. (A) Stage rotated until the crystallographic direction, cleavage trace 1 in this case, is parallel to the privileged direction of the polarizer—that is, to the N-S cross hair. (B) Rotation of the stage through angle T, the amount of rotation necessary to bring the gain to an extinction position.

Angle T is the difference between the stage readings for (A) and (B).

terminated by a crystal face. In either case, the extinction angle can be measured by aligning the observed crystallographic direction with a cross hair (Fig. 8–19A) and recording the angular position of the microscope stage as indicated at the index mark. The stage should then be rotated until the crystal plate is at extinction (Fig. 8–19B), its privileged directions now parallel the cross hairs; the stage's angular reading at the index mark is again noted. The difference between the reading for positions A and B equals T, the extinction angle. The extinction angle between cleavage trace 1 and the second privileged direction (ω) would be $(90 - T)$. This complementary relationship always pertains since the two privileged directions are always mutually perpendicular.

In uniaxial crystals, only two types of extinction are observed: (1) *parallel* and (2) *symmetrical*. The type depends upon whether the crystallographic directions involved are parallel (see *m* planes in Fig. 8–20A) or at an angle (see *u* planes in Fig. 8–20A) to the optic axis of the crystal. An elongated crystal lying on one of the *m* planes and bounded by another of them (Fig. 8–20B) shows parallel extinction, angle T equaling 0 degrees. On the

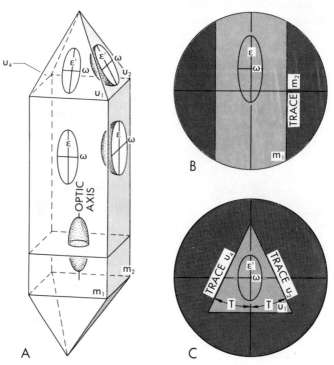

Fig. 8-20. (A) Intersections between the optic indicatrix and the prismatic faces (*m*) and pyramidal faces (*u*) of a tetragonal crystal. (B) View between crossed nicols of a crushed fragment of the crystal that is resting on one prism face and is bounded laterally by the traces of others. The angle between the trace of m_2 and the grain's ϵ privileged direction is 0 degrees; for this reason, the grain is said to have *parallel extinction*. (C) View between crossed nicols of a fragment at extinction that is resting on one pyramidal face and is bounded by the traces of two others. The angle T between its privileged direction and either trace u_2 or u_4 is equal. This grain thus possesses *symmetrical extinction*. For illustrative purposes the extinct crystals in (B) and (C) have been made slightly lighter than the background.

other hand, fragments of a crystal that cleaved parallel to a pyramidal set of faces—that is, parallel to faces labeled *u* in Fig. 8–20A—would likely rest on a *u* plane and be bounded by other *u* planes (Fig. 8–20C). The extinction angle between the ϵ' direction and cleavage trace u_4, would precisely equal the angle between this ϵ' direction and trace u_2. Thus arose the term "symmetrical extinction."

If a grain is elongated dimensionally such as that in Fig. 8–20B, two possibilities exist with respect to the orientation of the N and n directions to this elongation: (1) The N direction is parallel (or within 45 degrees) to the elongation (as for Fig. 8–20B), in which case the grain is said to be *length slow* or to possess a *positive elongation;* or (2) The n direction is parallel (or within 45 degrees) to the direction of grain elongation, in which case the grain is said to be *length fast* and to possess a *negative elongation.*

ABSORPTION AND PLEOCHROISM

Like isotropic materials (p. 16), uniaxial crystals may exhibit general absorption (to appear gray or opaque in transmitted light) or selective absorption (to appear colored). As with refractive index, absorption may differ according to vibration direction within a uniaxial crystal. Tourmaline, for example, is generally strongly absorbent for light vibrating parallel to an ω direction whereas for light vibrating parallel to its optic axis, the crystal is highly transparent (Fig. 8–21).

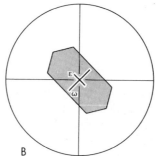

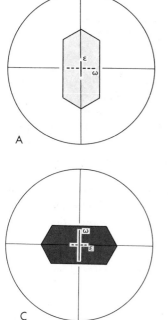

Fig. 8-21. Tourmaline, which shows strong general absorption for light vibrating parallel to ω. No analyzer, N-S polarizer only. The crystal changes from colorless (A), to gray (B), to black (C) as the microscope stage is rotated. For orientation (C) the crystal absorbs almost all of the N-S vibrating light that enters it.

For some minerals, either ϵ or ω, or both, may represent a direction of selective absorption. Moreover, both directions, if absorbent, need not absorb the same portion of the spectrum. In such cases a crystal plate, rotated within a beam of polarized light, is observed to transmit different colors according to whether the ϵ or ω vibration is parallel to the privileged direction of the polarizer (Fig. 8–22). This change in color with rotation in the polarized beam is called *pleochroism*.

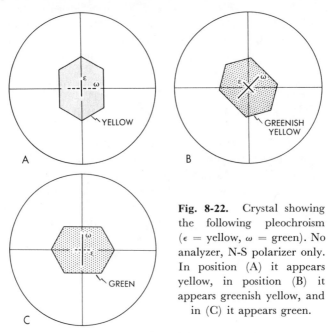

Fig. 8-22. Crystal showing the following pleochroism (ϵ = yellow, ω = green). No analyzer, N-S polarizer only. In position (A) it appears yellow, in position (B) it appears greenish yellow, and in (C) it appears green.

A crystal's coefficients of absorption* for light vibrating at 90 and 0 degrees to the optic axis—respectively symbolized as k_ω and k_ϵ—may vary for different wavelengths of light. Fig. 8–23A illustrates this variation for a crystal that exhibits a strong general absorption for light vibrating perpendicular to the optic axis but little absorption for light vibrating parallel to the optic axis. Fig. 8–23B similarly illustrates the variation in k_ω and k_ϵ for a crystal that selectively absorbs all wavelengths but green for light vibrating perpendicular to the optic axis and all wavelengths but yellow for light vibrating parallel to the optic axis. This second crystal thus has the following pleochroic formula: ω = green; ϵ = yellow. For light of a particular wavelength vibrating within this crystal at an angle θ to the optic axis, the absorption coefficient $k_{\epsilon'}$ varies between k_ϵ and k_ω for this wavelength, approaching k_ϵ as the value of θ decreases. For this reason, pleochroic colors may vary according to the orientation of the section within the crystal. For example, a plate cut nearly perpendicular to the optic axis—so that its ϵ' index differs little from ω—shows the following pleochroic colors if viewed in plane-polarized light: ϵ' = yellowish green; ω = green. A plate cut exactly perpendicular to the optic axis would appear green for all positions of the stage.

---

* See p. 17 for the definition of coefficient of absorption.

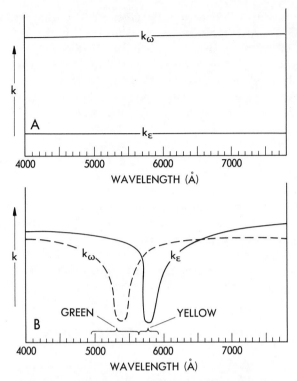

Fig. 8-23. Variation for different wavelengths of the absorption coefficients for light vibrating parallel to the optic axis (k_ε) and perpendicular to it (k_ω). (A) For the crystal in Fig. 8-21 for which light vibrating perpendicular to the optic axis was almost completely absorbed for all wavelengths and light vibrating parallel to the optic axis was almost completely transmitted. (B) For the crystal in Fig. 8-22 for which the pleochroism was ε = yellow and ω = green.

ABNORMAL INTERFERENCE COLORS

An abnormal interference color is one whose hue does not match any of the normal interference colors of the scale (Fig. 8–17). Abnormal colors indicate that the spectral composition of the light being transmitted by the analyzer (after its prior passage through the mineral) differs from that of crystal plates and grains (of comparable retardation values) of the majority of minerals. In uniaxial crystals abnormal interference colors are usually produced as the result of (1) strong selective absorption of a particular wavelength during its passage through the crystal; or (2) large variation in the value of $(\varepsilon - \omega)$ for different wavelengths of light, $(\varepsilon - \omega)$ sometimes decreasing to zero at a particular wavelength. By way of an example of the first possibility, consider a deep green crystal whose retardation is 200 mμ. Ordinarily, the spectral transmission by the analyzer for a retardation of 200 mμ is like the dashed curve in Fig. 7–8. For this crystal, however, as its green color testifies, a great part of the red wavelengths of light was absorbed during passage through the crystal. Consequently, a different spectral transmission curve results and, instead of the first-order white usually seen for a retardation of 200 mμ, a greenish interference color would be seen.

For a few minerals the coefficients of dispersion for ϵ and ω differ significantly in value and their birefringence ($\epsilon - \omega$) thus differs appreciably for different colors of light. In metatorbernite (see Fig. 8–15) the birefringence (and therefore the retardation of a given plate) is, for red light, double what it is for blue light. Moreover, wavelengths in the region of 5120 Å, for which metatorbernite is isotropic, will be completely extinguished at the analyzer (and therefore absent from the spectral composition of the interference colors) for all plates of metatorbernite, regardless of their thickness. Consequently, plates of metatorbernite yield abnormal interference colors between crossed nicols, all lacking the wavelengths around 5120Å.

MINERAL IDENTIFICATION

After ϵ and ω have been accurately measured for an unknown uniaxial mineral, the particular mineral species to which it belongs may be determined by consulting Figs. II-1 or II-2 in Appendix II (pp. 253 and 254). Details of this method of identification are given on pages 242–3.

NINE — BIAXIAL CRYSTALS

A biaxial crystal belongs to the orthorhombic, monoclinic, or triclinic system and possesses *three* significant indices of refraction, commonly symbolized as α, β, and γ. Of these, α and γ represent, respectively, the smallest and largest refractive indices exhibited by the crystal; β is intermediate in value between them. Less popular, alternative symbols for these indices are sometimes used (Table 9–1).

Table 9–1

REFRACTIVE INDEX SYMBOLISMS
(AND FREQUENCY OF USAGE IN PERCENT)

Smallest index	α	n_α	n_p	n_x	nX	X	n_a	N_x	n_1
Intermediate index	β	n_β	n_m	n_y	nY	Y	n_b	N_y	n_2
Largest index	γ	n_γ	n_g	n_z	nZ	Z	n_c	N_z	n_3

Frequency in 42 recent texts and standard works*

α	n_α	n_p	n_x	nX	X	n_a	N_x	n_1
47.6	26.1	9.5	11.9	2.4		2.4		

Frequency of use by 39 European scientists**

α	n_α	n_p	n_x	nX	X	n_a	N_x	n_1
71.8	7.7	15.4	5.1					

Frequency of use in 27 articles in 1956 *American Mineralogist*

α	n_α	n_p	n_x	nX	X	n_a	N_x	n_1
74.1			3.7	14.8		3.7	3.7	

* Source: Survey report by Nomenclature Committee of Mineralogical Society of America (1957).
** Source: Report by Earl Ingerson on 1954 visit to Denmark, Finland, France, Germany, Great Britain, Holland, Iceland, Norway, Sweden, and Switzerland.

BIAXIAL INDICATRIX

As in uniaxial crystals, the index of refraction for monochromatic light varies with its vibration direction in biaxial crystals. Within such crystals two unique, mutually perpendicular directions may be located for which, and only for which, the crystal respectively exhibits its greatest and least refractive indices, γ and α. For a third direction, at right angles to these two, the crystal exhibits the refractive index β. These three directions, commonly symbolized as X, Y, and Z, are called the three *principal vibration axes*. The correspondence between these directions and the refractive indices that the crystals exhibit for light vibrating parallel to them is always as illustrated in Fig. 9–1A.

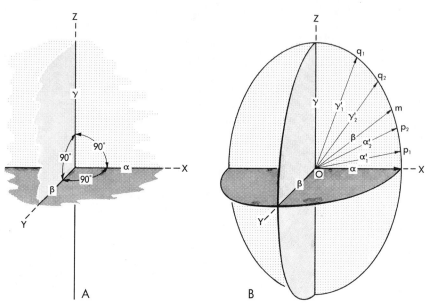

Fig. 9-1. (A) The three, mutually perpendicular, principal vibration axes, X, Y, and Z, and the common symbols for the indices of refraction of a biaxial mineral for light vibrating parallel to them. (B) Elliptical distribution of the index of refraction (as shown by the vector lengths) for light vibrating parallel to op_1, op_2, om, oq_2, oq_1 within the ZX plane.

For vibration directions not coinciding with X, Y, or Z, the corresponding refractive indices will generally be between α and β or between β and γ in value. The former index is symbolized as α', the latter as γ'. Consequently, the following relationship always holds true : $\alpha < \alpha' < \beta < \gamma' < \gamma$.

Consider now a few of the random vibrations within the XZ plane (stippled in Fig. 9–1B). The varying lengths of these radii, drawn to represent the crystal's refractive index for that vibration, define an ellipse whose semiminor axis is α and whose semimajor axis is γ. Note that the length *om* is exactly equal to β. Similarly, the lengths of random vibrations in the lightly shaded YZ plane also define an ellipse (semiminor axis, β; semimajor axis, γ), and those lying in the deeply shaded XY plane, a third ellipse (semiminor axis, α; semimajor axis, β). Fig. 9–1B thus illustrates the basic framework of the biaxial indicatrix.

Countless planes hinging on the Z axis (in addition to the XZ and YZ planes of Fig. 9–1B) could be drawn; in Fig. 9–2A unshaded planes 1, 2, and 3 are examples. As is demonstrated in detail for plane 2, the crystal's refractive indices vary with vibration direction in each such plane. The vectors representing the crystal's indices for light vibrating parallel to them vary in length like radii of an ellipse. The semiaxes of this ellipse for plane 1 are γ and α_1'; for plane 2, γ and α_2'; and for plane 3, γ and α_3'. In each case the semiminor axis of the ellipse—that is, α_1', α_2', or α_3'—is also a radius of the deeply shaded $\alpha\beta$ ellipse. Since $\alpha' < \beta < \gamma$, in each of these ellipses hinging on Z there exist radii equal to β in length. All these β radii lie

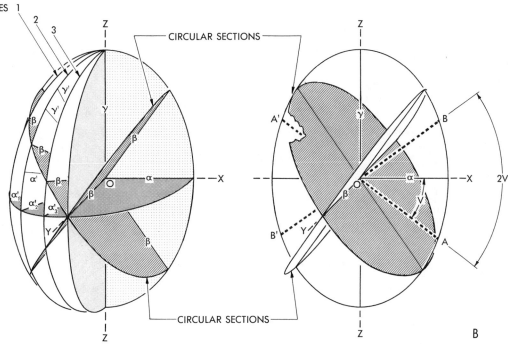

Fig. 9-2. (A) Variation of refractive index in a biaxial crystal as shown by ellipses (whose radii are proportional in length to the crystal's refractive index for light vibrating parallel to them). Planes 1, 2, and 3 are typical of the numerous ellipses that could be drawn to hinge on ZZ. All such ellipses contain one radius equal to β in length, these radii lying in the same planes (ruled). These planes, the circular sections, intersect in principal axis Y. (B) The biaxial indicatrix; that is, simply an imaginary three-dimensional ellipsoid whose radii are proportional to the crystal's refractive indices for light vibrating parallel to them. The normals to the circular sections, AA' and BB', are the two optic axes.

in the same plane and define two circles (of radius β), each closely ruled in Fig. 9–2A.

Nomenclature

The shape of the biaxial indicatrix has now emerged (Fig. 9–2B). It is a three-dimensional ellipsoid, all central sections of which are ellipses except for two. These two, one unshaded and one closely ruled in Fig. 9–2B, are the *circular sections*, the radius of each being β. The two normals to these circular sections, AA' and BB', are called the (primary) *optic axes*, a term often abbreviated as O. A. Since there are two such optic axes (each comparable with the unique optic axis of the uniaxial indicatrix), the origin of the name biaxial indicatrix becomes obvious.* The two optic axes always lie within the XZ ($\alpha\gamma$) plane, which is therefore called the *optic plane*. The principal vibration direction Y, since it is always normal to the optic plane, is called the *optic normal* (abbreviated O. N.). The acute angle between the two optic axes is called $2V$ or simply the *optic axial angle*. The obtuse axial angle—that is, angle BOA'—is always supplementary to $2V$.

The principal vibration axis that bisects the (acute) optic axial angle (that is, X in Fig. 9–2B) is called the *acute bisectrix* (abbreviated A. B. or Bx_a). The principal vibration axis which bisects the obtuse axial angle is called the *obtuse bisectrix* (abbreviated O. B. or Bx_o). In Fig. 9–2A, X (α) is shown to be the acute bisectrix whereas Z (γ) is the obtuse bisectrix. For many minerals this will be reversed; that is, Z will be the acute bisectrix and X will be the obtuse bisectrix. In biaxial minerals the optic sign depends upon this relationship. Biaxial ($+$) crystals are defined to be those for which Z (γ) is the acute bisectrix; biaxial ($-$) crystals are those for which X (α) is the acute bisectrix. For ($+$) minerals, β is closer in value to α than to γ. For ($-$) minerals β is generally closer to γ than to α but rare exceptions may occur, particularly for minerals in which $2V$ is near 90 degrees. The biaxial indicatrix of a mineral is usually drawn with Z vertical.

Equations for the Biaxial Indicatrix

The equation for the biaxial optical indicatrix is generally given as

$$\frac{X^2}{\alpha^2} + \frac{Y^2}{\beta^2} + \frac{Z^2}{\gamma^2} = 1 \qquad \text{(Eq. 9–1)}$$

the equation for a triaxial ellipsoid in rectangular coordinates.

From the polar equations for an ellipse, an equation for the biaxial indicatrix may be developed that uses stereographic coordinates and is therefore well suited for the stereographic solution of optical problems. Let OP be a randomly directed radius of a biaxial indicatrix (Fig. 9–3). Define ρ (rho) as its angle with principal axis Z, and ϕ (phi) as the angle

---

* In geometrical terminology the biaxial indicatrix is a triaxial ellipsoid in reference not to the optic axes but to its three, mutually perpendicular, principal axes, X, Y, and Z.

between X and the unshaded plane which contains OP and Z. Thus ϕ and ρ represent the stereographic coordinates that serve to orient line OP in the indicatrix; such stereographic coordinates are discussed in detail in

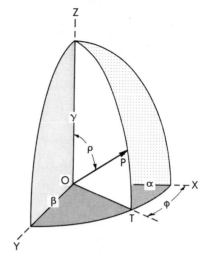

Fig. 9-3. Stereographic coordinates, ϕ and ρ, of OP, a randomly directed radius of the biaxial indicatrix whose principal axes X, Y, and Z are as shown. Z is vertical as preferred for biaxial indicatrices.

Chapter 11. The length of OP, and therefore the crystal's index γ'_{op} (or α'_{op}, if computation discloses it to be less than β) for light vibrating parallel to OP, may be calculated from the following equation

$$\gamma'_{op}(\text{or } \alpha'_{op}) = \frac{1}{\sqrt{\dfrac{\sin^2 \rho \cos^2 \phi}{\alpha^2} + \dfrac{\sin^2 \rho \sin^2 \phi}{\beta^2} + \dfrac{\cos^2 \rho}{\gamma^2}}} \qquad \text{(Eq. 9–2)}$$

Variation and Calculation of 2V

The angle between the two optic axes, if denoted as $2V$, is by definition the acute angle between them. An increasingly popular convention, however, uses the symbols $2V_x$ or $2V_z$ to denote the angle between the optic axes according to whether it is measured across the X direction or across the Z direction (Fig. 9–4). The angles so defined may then exceed 90 degrees. The sum of $2V_x$ plus $2V_z$ for a given crystal—for example, either crystal in Fig. 9–4—always equals 180 degrees. Note that the statement that a biaxial crystal is optically negative in sign with $2V = 37°$ may be succinctly restated in this new symbolism as $2V_x = 37°$ or, alternatively, as $2V_z = 143°$. Either of these latter statements implies that X bisects the acute angle between the optic axes and that Z bisects the obtuse angle between them.

The value of the optic axial angle, $2V$, depends upon the extent to which β is closer to α than to γ in value, or vice versa. If, for example, β differs slightly from α but considerably from γ (see Fig. 9–5A), then Om (the radius in the $\alpha\gamma$ ellipse equal to β) is at a small angle to the unshaded $\alpha\beta$ ellipse. Thus $2V$ is very small, and γ is the acute bisectrix. Consider now

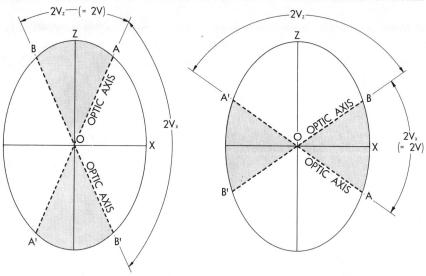

Fig. 9-4. Relationship between $2V_x$ and $2V_z$. Note that $2V$ equals $2V_x$ or $2V_z$, whichever is smaller. The dashed lines represent the optic axes.

a second mineral possessing the same values of α and γ but a larger value of β (Fig. 9–5B). In this case, Om is at a larger angle to the $\alpha\beta$ ellipse. Thus V (and consequently $2V$) is larger than for the first mineral.

For a third mineral, again assume the same values of α and γ but this time assume β to be approximately (but not exactly) halfway between α and γ (Fig. 9–5C). For this mineral, V is 45 degrees and $2V$ is 90 degrees. As a consequence, neither α nor γ can be properly called the acute bisectrix, since the angle $2V$ is a 90-degree angle rather than acute. Hence this third mineral is neither $(+)$ nor $(-)$. For a fourth mineral, α and γ are as before but now β is closer to γ in value (Fig. 9–5D). Consequently, Om (the radius in the $\alpha\gamma$ ellipse equal to β) is close to the Z axis. As a result, the angle between OX and OA (the optic axis) is now acute; OX has now become the acute bisectrix. This fourth mineral is therefore $(-)$ in sign. Note that if β becomes equal to α in value, the mineral becomes uniaxial positive; if β equals γ, the mineral becomes uniaxial negative.

The angle $2V$ can be readily calculated if the values of α, β, and γ are known. The relationship is

$$\cos V_z = \frac{\alpha}{\beta} \sqrt{\frac{(\gamma + \beta)\ (\gamma - \beta)}{(\gamma + \alpha)\ (\gamma - \alpha)}} \qquad \text{(Eq. 9–3)}$$

where V_z is the angle between the optic axis and the Z direction.* For

---

* Eq. 9–3 can be derived from Eq. 9–2. Simply set $\phi = 0°$ (which reduces Eq. 9–2 to the polar equation for the $\alpha\gamma$ ellipse) and substitute β for γ'_{op}. The particular value of ρ that satisfies the resultant equation is then the complement of V (this complementary relationship is well shown in Figs. 9–5 C and D where angle mOZ is ρ). Algebraic manipulation and use of the identity that $\cos V_z = \sin \rho$ ultimately results in Eq. 9–3.

biaxial negative minerals V_x may be preferred. If V_z has been calculated, V_x can be readily determined since

$$V_z + V_x = 90° \qquad \text{(Eq. 9–4)}$$

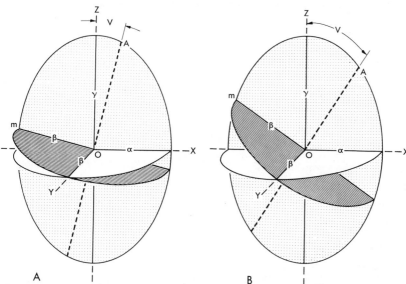

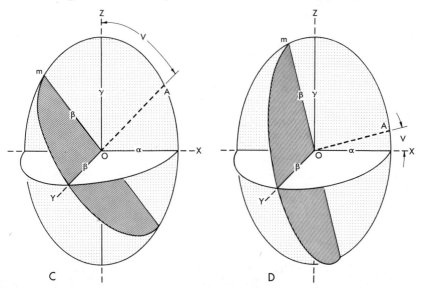

Fig. 9-5. The variation of $2V$ in four crystals all possessing equal values of α and γ but, from (A) to (D), increasing values of β. For simplicity only one circular section (closely ruled) and one optic axis (OA) are drawn. Note the attendant increase in $2V$ from (A) to (C). With increase in β beyond that in (C), the optic sign becomes negative. Note that V_z increases continuously from (A) to (D).

If V_z is a value greater than 45 degrees, it is obvious that the crystal is actually biaxial negative. The value of $2V$ may be quickly calculated, usually to within 2 degrees, by use of the following approximate relation: $\cos 2V = 1 - 2\,[(\beta - \alpha)/(\gamma - \alpha)]$. Negative values for $\cos 2V$ indicate the crystal to be negative in optic sign; positive values indicate optically positive crystals.

Eq. 9–3 indicates that, given any three of the four variables α, β, γ,

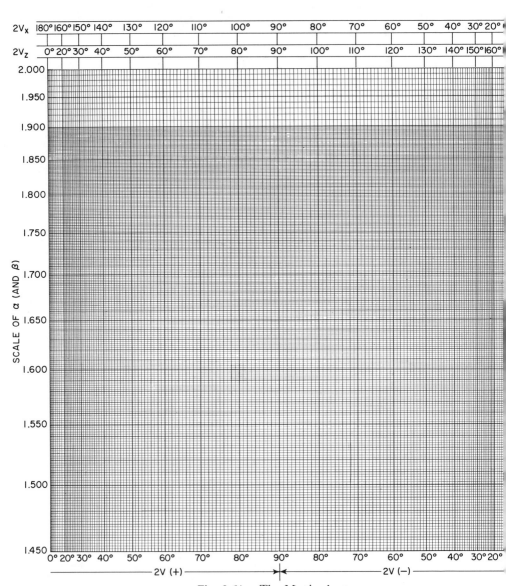

Fig. 9-6A. The Mertie chart.

and V, the fourth may be calculated. A nomogram (Fig. 9–6A) devised by
J. B. Mertie (1942, p. 538) eliminates the need for any calculations. Fig. 9–6B
illustrates its use. Simply place a transparent straight-edge to indicate the
value of α on the left edge of the chart (point 1) and the value of γ on its
right edge (point 3). Locate the point on the line joining 1 and 3 for which
the refractive index equals β (this is point 2 in Fig. 9–6B). A vertical line
through point 2 (marked by an arrow in Fig. 9–6B) indicates the value of $2V$

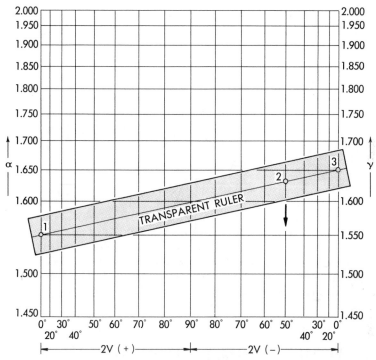

Fig. 9-6B. Method of determining $2V$ with the Mertie chart (Fig. 9-6A)
from the values of α (point 1), β (point 2), and γ (point 3). The transparent
straight edge has a straight line ruled along the center of its bottom
surface.

on Fig. 9–6A's upper or lower scale. Thus, in Fig. 9–6B, $\alpha = 1.550$, $\beta = 1.630$,
$\gamma = 1.650$, and thereby $2V$ is determined to be 50 degrees, the mineral's
sign being (−). In a similar manner, if $2V$, the optic sign, and two of the
three indices α, β, and γ are known, points 1, 2, and 3 can be located. (They
are always collinear.) Consequently, the value of the third index can be
determined from the ordinate value of point 1, 2, or 3 (depending upon
whether α, β, or γ represents the unknown value).

The angle $2V$ represents the angle between the two optic axes within
the crystal. The angle in air between light rays that traveled along the two
optic axes while in the crystal is referred to as $2E$ (Fig. 9–7). As we shall
later discuss in detail, the refractive indices for rays traveling along the

optic axes in biaxial crystals is β. Thus, applying Snell's law, the relationship between E and V is seen to be

$$\beta \sin V = \sin E \qquad \text{(Eq. 9-5)}$$

The Mertie article (1942) also includes a chart, similar to Fig. 9–6A, for graphic determination of E. For our purposes, however, $2V$ can be graphically determined by using Fig. 9–6A, then converted to $2E$ by using a slide-rule solution of Eq. 9–5.

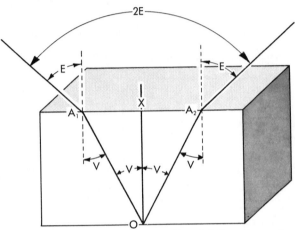

Fig. 9-7. Relationship between $2E$ and $2V$ as shown by the refraction in air of two rays that, while in the crystal, traveled along the optic axes, OA_1 and OA_2. If these two rays emerge in a medium of different index from air, the angle between them is called $2H$ rather than $2E$.

GEOMETRIC RELATIONSHIPS BETWEEN WAVE NORMALS, VIBRATION DIRECTIONS, AND RAY PATHS

Indicatrix Theory

For a random wave normal in a biaxial indicatrix (for example, OW in Fig. 9–8), the associated privileged directions lie, by definition, in a plane perpendicular to the wave normal. When this plane is extended outward from the indicatrix center, it intersects the indicatrix in an ellipse (stippled in Fig. 9–8). The semiminor and semimajor axes of this ellipse (for example, On and ON in Fig. 9–8) represent the only two privileged directions that can be associated with the given wave normal. The crystal's indices for light vibrating parallel to On or to ON are α' or γ', respectively.

Only one ray path can be linked with a given wave normal and one of its two associated privileged directions; all three (path, normal, and privileged direction) always lie within the same plane. For example, ray path OR_2 (Fig. 9–8) is the only one that can be associated with wave normal OW and privileged direction ON. It is readily located since (1) it lies in the same ellipse as do OW and ON—that is, in the lightly shaded ellipse in Fig. 9–8—and (2) it is the radius conjugate to ON in this ellipse. Similarly, ray path OR_1 is located in the same ellipse as OW and On—that is, in the more heavily shaded ellipse in Fig. 9–8—and is conjugate to On.

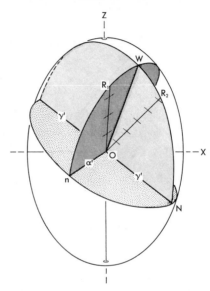

Fig. 9-8. A given wave normal direction, OW, is shown in respect to ON and On, the only two vibration or privileged directions possibly associable with it; angles WON and WOn are therefore 90 degree angles. Note that ON and On are the major and minor axes of the ellipse (stippled) obtained by passing a plane (perpendicular to OW) through the indicatrix. Each of these vibration directions, in conjunction with OW, defines a plane that also cuts the indicatrix in an ellipse (shaded). Ray paths OR_1 or OR_2, the only two ray paths associable with wave normal OW, lie within these ellipses, being radii of the ellipse conjugate to the vibration direction.

Biot-Fresnel Rule

Biot (1820) and Fresnel (1827) also formulated a rule that permits determination of the two privileged directions associable with a given wave

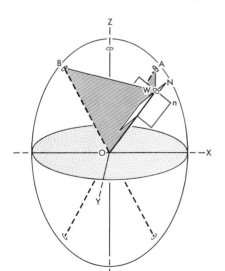

Fig. 9-9A. The vibration directions, WN and Wn, that are associable with wave normal OW, as obtained by the Biot-Fresnel rule. (OA and OB represent the two optic axes.)

normal direction. The rule is of great practical use since it is easily adapted to the stereographic solution of optical problems. Given the wave normal direction OW and the two optic axes OA and OB in the crystal (Fig. 9–9A), the vibration directions WN and Wn associated with wave normal OW may be located as follows: Construct a plane containing OA and OW and another containing OB and OW (in Fig. 9–9A, each is finely ruled). The vibration directions, WN and Wn, lie in the unshaded planes that bisect the angle between these finely ruled planes. Since vibration directions WN and Wn are necessarily perpendicular to OW as well, their precise direction in the crystal is known.

Vibration Directions Associable with a Given Ray Path

Given a ray path, the two vibration directions associable with it are difficult to determine unless the path lies within a plane parallel to two of the three principal vibration axes, X, Y, and Z (such planes are here termed principal planes for the biaxial indicatrix). For rays traveling within principal planes as thus defined, the relationship between a ray path and its two associable vibration directions is analogous to the uniaxial case (see p. 91). For example, OR_1, a ray path within the ZY principal plane in Fig. 9–9B, may represent a ray vibrating perpendicular to this principal plane (that is, parallel to OX) or a ray vibrating within this principal plane along a direction conjugate to its path (that is, WW in Fig. 9–9B). Thus a ray direction within a principal plane is associable with (1) a vibration direction perpendicular to this principal plane (in which case it acts as an O ray) or (2) a vibration direction within this principal plane that is conjugate to the ray path (in which case it acts as an E ray, its path not obeying Snell's law). Ray OR_2 in Fig. 9–9B serves as an additional example.

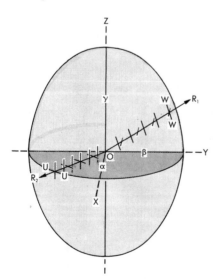

Fig. 9-9B The vibration directions associable with a ray path (OR_1) within a principal plane (ZY). One is perpendicular to this plane (that is, is parallel to OX); the second (WW) lies within this plane, being the conjugate direction to the ray path. Ray path OR_2 is an additional example of the principle.

Examples for Normal Incidence

As with uniaxial crystals, normal incidence constitutes for biaxial crystals the simplest yet most important type of incidence. For normal incidence, the revised Snell's law indicates that, since the incident wave normal direction is perpendicular to the crystal face or surface, the wave normal of the refracted light must also be (that is, $\measuredangle\, i = \measuredangle\, r = 0°$). Thus the two privileged directions associable with the refracted wave normal, since they must always be perpendicular to this wave normal, necessarily lie parallel to the crystal plane upon which normal incidence occurs. More specifically, these two privileged directions represent the major and minor axes of the ellipse formed by the intersection of the crystal plane with the indicatrix, the plane always being assumed to pass through the indicatrix center (see p. 81).

The dimensions of the major and minor axes of the ellipse of intersection depend upon the angular attitude of the plane to the X, Y, and Z principal vibration axes of the indicatrix. Thus, according to whether they are parallel to two, one, or none of these principal vibration axes, crystal planes may be classified (Table 9–2) into three groups: (1) principal planes, (2) semi-

Table 9–2

TYPES OF PLANES THROUGH THE BIAXIAL INDICATRIX

Principal axes to which crystal plane is parallel	Convenient optical symbol for plane	Dimensions of ellipse of intersection between plane and indicatrix	
		minor axis	major axis
(1) Principal planes			
X, Y	XY	α	β
X, Z	XZ	α	γ
Y, Z	YZ	β	γ
(2) Semirandom planes			
X	XZ'	α	γ'
Y	YX'	α'	β
Y	YZ'	β	γ'
Y (plus circular section)	YC	circle of radius β	
Z	ZX'	α'	γ
(3) Random planes			
None	$X'Z'$	α'	γ'

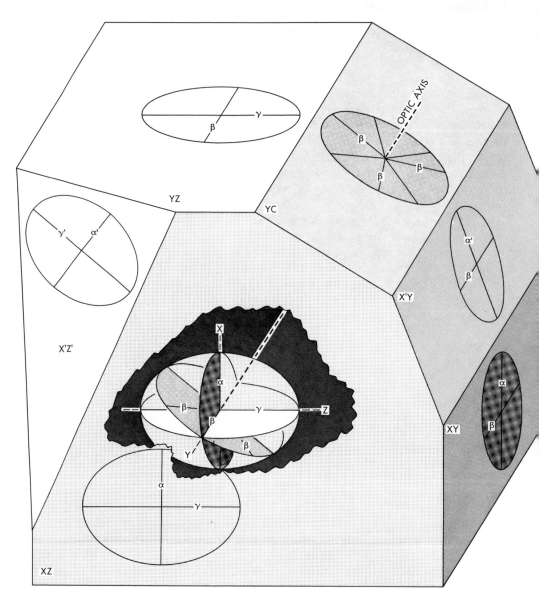

Fig. 9-10. The biaxial indicatrix in skeletal form is shown at the center of the crystal. Its intersection with different faces of the crystal, if it were moved translationally until its center fell on a crystal face, is shown for several faces. The major and minor axes of these ellipses of intersection represent the privileged directions of the crystal for normal incidence on that face. For face *YC*, since it intersects the indicatrix in a circle, innumerable privileged directions exist. The crystal's refractive indices for the privileged directions are also shown. See Table 9-2 for details of symbolism used for the faces.

random planes, or (3) random planes. An individual plane may be conveniently symbolized by two letters denoting the directions in the indicatrix to which the plane is parallel. The letters X' and Z', for example, represent directions within the indicatrix that are on opposite sides of the circular section. Direction X' is on the same side as X; Z' is on the same side as Z. The symbol YC has been used to indicate a circular section, since a plane parallel to the circular section is always parallel to the Y principal vibration axis as well. Examples of the different types of faces are illustrated in Fig. 9–10.

Random Planes. The geometric principles discussed earlier (pp. 160–1) govern the location of the refracted ray paths for normal incidence. Fig. 9–11A illustrates the general case for normal incidence, that upon a random plane. The refracted wave normal OW is associable with either vibration direction OZ' or OX'. Only ray path OR_1 is associable with wave normal OW and vibration direction OZ'; all three lie within the same plane (heavily shaded in Fig. 9–11A). This plane intersects the indicatrix in an

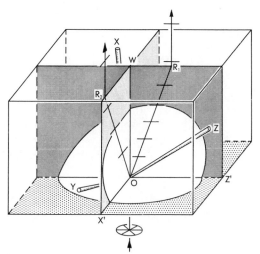

Fig. 9-11A. The ray paths developed within a biaxial crystal as the result of normal incidence of an unpolarized ray upon a random section $(X'Z')$. The ray-containing planes defined by wave normal OW and vibrations OZ' or OX' are shaded dark and light, respectively, each being a random section. Deviations of the ray paths OR_1 and OR_2 from the wave normal OW have been exaggerated for illustrative purposes.

ellipse for which OR_1 and OZ' are conjugate radii. Similarly, ray path OR_2 is the only one associable with wave normal OW and vibration direction OX'. Consequently, if unpolarized light is normally incident upon a random section, two rays, OR_1 and OR_2, result. Both act as extraordinary rays, neither coinciding with their common wave normal direction OW. If the incident light is polarized, ray OR_1, ray OR_2, or both OR_1 and OR_2 result, according to whether the direction of light vibration is parallel to OZ', to OX', or to neither.

Semirandom Planes. Fig. 9–11B illustrates a more special case, normal incidence upon a semirandom section. The situation for ray OR_2 is as before. However, ray OR_1 (the only ray associable with wave normal OW and vibration direction OZ) now coincides with OW in direction (since OW is the radius conjugate to OZ for the ellipse of intersection between heavily shaded plane R_1OZ and the indicatrix). Note that the lightly shaded plane is principal section XY of the indicatrix. It contains principal axes OY and OX and is perpendicular to OZ (and therefore to the stippled, semi-random plane upon which the unpolarized ray is normally incident). Rays OR_2 and OR_1 both lie in principal plane XY and bear the same relationship to it as do the E and O rays to a principal plane of a uniaxial crystal (p. 91, points 7–10).

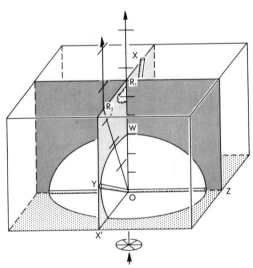

Fig. 9-11B. Rays resulting from the normal incidence of an unpolarized ray on a semirandom section $X'Z$. The ray-containing plane defined by wave normal OW and vibration OX' is a principal section.

The generalization already discussed (see p. 162) again becomes apparent: A ray traveling within a principal plane of a biaxial crystal vibrates either perpendicular to this plane (for example, ray OR_1 in Fig. 9–11B) or within this plane along a direction conjugate to its path (for example, OR_2). For the first case, ray path and vibration direction are mutually perpendicular and the ray acts like an ordinary ray. For the second, the ray acts like an extraordinary ray, only its wave normal following Snell's law.

Principal Plane. Light normally incident upon a principal section of a biaxial crystal is the most specialized case of all. Fig. 9–11C illustrates that, for unpolarized incident light, two rays are produced, each vibrating parallel to a principal direction of the indicatrix and each coinciding in path with wave normal OW.

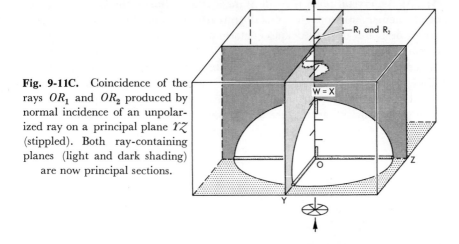

Fig. 9-11C. Coincidence of the rays OR_1 and OR_2 produced by normal incidence of an unpolarized ray on a principal plane YZ (stippled). Both ray-containing planes (light and dark shading) are now principal sections.

Circular Section; Internal Conical Refraction. A thin pencil of unpolarized light, if normally incident upon a circular section of a biaxial crystal, becomes a hollow cone of light within the crystal and a hollow cylinder of light after emergence (Fig. 9–12A). This phenomenon, known as internal conical refraction, is readily explained by indicatrix theory. As previously discussed (p. 68), unpolarized light normally incident upon a circular section is not constrained to vibrate parallel to any particular direction after entering the crystal. Thus, both before and after entry into the crystal, the unpolarized light vibrates in innumerable directions within the circular section; a few of these possible vibration directions are illustrated in Fig. 9–12A as radii labeled "β." After entry into the crystal, all light vibrating parallel to these various radii possesses the common wave normal direction OW (a consequence of normal incidence). Each of these different vibration directions, in conjunction with wave normal OW, gives rise to a different ray path (see p. 160). Thus, light vibrating parallel to OY in Fig. 9–12B produces ray 1 (ray 1 is the radius conjugate to β_1 in the lightly shaded ellipse YOW). Similarly, light vibrating parallel to β_2 produces ray 2 (β_2 and ray 2 are conjugate radii lying within the dotted ellipse); light vibrating parallel to β_3 produces ray 3 (β_3 and ray 3 are conjugate radii of heavily shaded ellipse WOX').

If, as for β_3, ray paths were determined for the numerous other vibration directions between β_2 and β_1, the conical distribution of these ray paths (with respect to their common wave normal OW) becomes apparent. Each ray path on the cone is thus the result of a different vibration direction; Fig. 9–12C, a slightly enlarged, aerial view of the cone of rays, shows the vibration direction for several rays at their points of emergence from the crystal. Note their orientation with respect to the trace of the optic plane. Rays 1 and 2 traveled in this plane, ray 1 acting like a uniaxial O ray, ray 2 like an E ray.

As the crystal plate is made thinner, the radius of the hollow cylinder of light emergent from the crystal decreases, becoming minimal for crystal grains or plates of the thicknesses normally dealt with in optical crystallography. These plates therefore, act much like the circular sections of uniaxial minerals (which are incapable of exhibiting internal conical refraction). Thus an unpolarized ray of normally incident light remains essentially unpolarized during transmission by these plates whereas a polarized ray retains the same plane of vibration even after entry into the crystal.

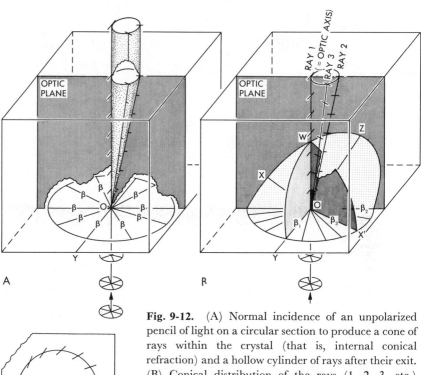

Fig. 9-12. (A) Normal incidence of an unpolarized pencil of light on a circular section to produce a cone of rays within the crystal (that is, internal conical refraction) and a hollow cylinder of rays after their exit. (B) Conical distribution of the rays (1, 2, 3, etc.) associated with different vibrations in the circular section (β_1, β_2, β_3, etc.) but with the common wave normal OW. (C) Slightly enlarged view of rays emerging on the crystal's upper surface, showing the vibration directions at a few points of emergence.

Oblique Incidence

Only special cases of oblique incidence will be discussed here—namely, oblique incidence for which the incident ray is parallel to a principal plane of the indicatrix. Fig. 9–13 is an example. The unpolarized ray UO is obliquely incident upon the crystal's undersurface, which is a YX' semirandom plane. The plane of incidence is therefore the XZ principal plane (whose dashed extension is drawn below the crystal in Fig. 9–13 to emphasize this). Two refracted rays result, both paths lying within this principal plane. One of these rays, OR_1, vibrates perpendicular to the principal plane containing its path. It thus acts as an O ray. In the example this ray vibrates parallel to OY and is associated with the refractive index β. Its path, since it is perpendicular to its vibration directions (and thus coincides with its wave normal), follows Snell's law; thus

$$\sin r_1 = \frac{n_i}{\beta} \sin i$$

where n_i is the refractive index of the medium surrounding the crystal; i, the angle of incidence; and r_1, the angle of refraction for ray OR_1 (Fig. 9–13).

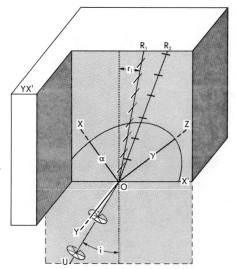

Fig. 9-13. Oblique incidence of UO, an unpolarized ray traveling parallel to the XZ plane on a semirandom plane YX'. A section of the crystal has been cut away for illustrative purposes. The vibrations for refracted ray OR_1 are parallel to OY; those for ray OR_2 lie within the XZ plane.

Ray path OR_2 is more difficult to determine. It acts as an E ray, its path and vibration direction not being mutually perpendicular. Both path and vibration direction lie within the principal plane to which the incident ray was parallel (the lightly shaded plane in Fig. 9–13). The path, vibration direction, and refractive index associated with this E ray may be determined by adapting the same graphical construction method as described on page 88 (steps 1 through 8).

BIAXIAL INTERFERENCE FIGURES

To be coherent—that is, capable of interfering with each other—two light rays must (1) originate from the same incident beam of polarized light upon its entrance into an anisotropic crystal and (2) travel along the same path after emergence from the crystal. If a biaxial crystal is illuminated by a solid cone of polarized light, closely adjacent rays within the cone may be considered to travel parallel to each other. Let a and b in Fig. 9–14 represent

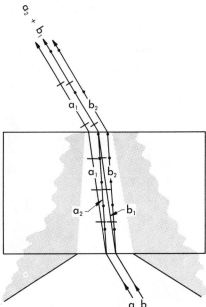

Fig. 9-14. Cross-sectional view of two closely adjacent rays a and b of the cone of illumination, their interspacing greatly exaggerated for illustrative purposes. Rays a and b respectively break up into rays a_1 and a_2 and b_1 and b_2 after entering an anisotropic crystal. Rays a_2 and b_1 emerge from the same point on the crystal and subsequently travel the same path in air, thus interfering with each other. Crystal thickness is greatly exaggerated.

two such rays, the spacing between them being measurable in angstroms but, for purposes of illustration, being greatly exaggerated. Upon entering the crystal, incident rays a and b break up into two rays, each vibrating in a different direction and each traveling with different speed; passage through the crystal retards ray a_2 with respect to ray a_1, and ray b_2 with respect to b_1. Rays such as a_2 and b_1, each from a different parent ray, emerge from the same point on the crystal and, since parent rays a and b were essentially parallel prior to entry into the crystal, a_2 and b_1 travel the same path in air, thus to interfere.

We will purposely make an incorrect assumption concerning Fig. 9–15A and in the discussion that follows—namely, that the two rays that mutually interfere after emergence have also traveled the same path while within the crystal. This of course is not true; ray paths a_2 and b_1, for example, do not coincide within the crystal of Fig. 9–14. However, within crystals of moderate birefringence the angle between ray paths such as a_2 and b_1 is generally small—that is about 1 to 3 degrees; thus our assumption is not too far from the truth and considerably simplifies the ensuing discussion and illustration.

Acute Bisectrix-centered Figures

Assume that a cone of light is focused with its apex at O on the bottom surface of a crystal (Fig. 9–15A). Upward from O, therefore, numerous rays diverge—some traveling as fast rays, others as slow rays, depending upon their vibration directions in the crystal. Of these multitudinous ray paths, consider OR, OP, OQ, OX, OS, OA, OT, and OU each to represent the path of a slow and a fast ray—their 1 to 3 degree angular separation being ignored—that mutually interfere after emergence from the crystal. The vibration directions for the slow and the fast ray along each such path are drawn at the point where the path intersects the indicatrix and again, this time orthographically projected, on the crystal's upper surface at the point of the ray's emergence. Since OX, OS, OA, OT, and OU all lie within the dotted principal plane ZX, one ray along each vibrated perpendicular to this plane while in the crystal and therefore is associated with a refractive index β. The second ray traveling along each of these paths vibrates within plane ZX but tangent to the indicatrix. Thus the refractive index for the second ray is γ for path OX, γ' for path OS, but α' for OT and a still smaller value of α' for path OU. For path OA, which coincides with the optic axis, the rays may vibrate parallel to any radius of the circular section; hence the only index associable with the ray along OA is β.

Similarly, OQ and OP in Fig. 9–15A lie within the lightly shaded XY principal plane. Therefore, one of the rays traveling along each vibrates parallel to OZ, the perpendicular to this plane; the index of this ray therefore corresponds to γ. The second ray along each vibrates within this plane but tangentially to the indicatrix. The refractive index for this second ray is therefore α' for OQ and a still smaller value of α' for OP. Path OR represents the general case since it does not lie within a principal plane. The fast and the slow ray traveling along it are associable with vibration directions corresponding to indices of α' and γ', respectively.

The retardation between the slow and fast rays when they emerge from the crystal's upper surface in Fig. 9–15A depends upon whether they traveled straight through the crystal (as along OX) or along a path at an angle θ to OX. The retardation between the slow and the fast ray emerging at OX equals $(\gamma - \beta)t$ whereas between the two rays assumed to travel along OS the retardation approximately equals $(\gamma' - \beta)(t/\cos\theta)$. Note that the retardation decreases to zero for slow-fast ray pairs traveling along paths within the fan between OX and OU as path OA is approached. This is because the index (γ' or α') associated with one ray of the pair approaches the value of the index associated with the remaining member of the pair, as a path becomes parallel to an optic axis. For paths making increasing angles to OA, the rate of increase in retardation is greater for those in the fan between OA and OU than in the fan between OA and OX, since path distance increases from OA toward OU whereas it decreases from OA toward OX. Retardation increases at the greatest rate in those paths lying outward from OX toward OP. For such paths the retardation upon emergence approximately equals $(\gamma - \alpha')(t/\cos\theta)$, the value of $(\gamma - \alpha')$ approaching that of $(\gamma - \alpha)$ as θ becomes large. For the numerous random ray paths that do not lie within principal planes—OR being an example—the retardation increases as their angle to the closest optic axis increases.

Surfaces of Equal Retardation; Isochromes. The surfaces joining all adjacent rays that emerge from the crystal with equal retardation are

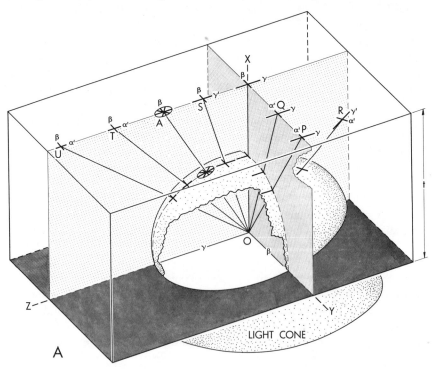

Fig. 9-15A. Relationship of selected ray paths OP, OQ, OR, OX, OS, OA, OT, and OU to their associated privileged directions, the latter shown as tangents to the indicatrix (or projected onto the crystal's upper surface at the ray's point of emergence).

known as *surfaces of equal retardation* or as Bertin's surfaces. In Fig. 9–15B they have been drawn only for retardations of 550, 1100, 1650, 2200, and 2750 mμ. The intersection between these surfaces and a horizontal plane parallel to the crystal's upper surface marks the pattern of isochromes visible in a biaxial interference figure in which the rays that traveled along the acute bisectrix emerge at the center of the field of view; this type of interference figure is thus called a centered acute bisectrix figure. If white light is the illuminant, these intersections are marked by first-, second-, third-, fourth-, and fifth-order red isochromes. If monochromatic light of wavelength 550 mμ is substituted as the illuminant, black areas of extinction appear where the red isochromes had been. Approximately midway between these black patterns, bright areas of light of wavelength 550 mμ (green) would be then seen.

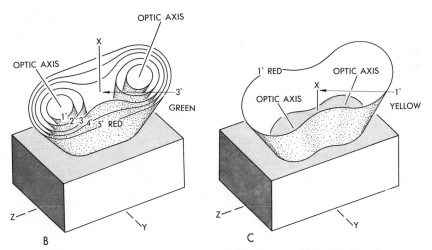

Fig. 9-15B-C.   (B) Surfaces of equal retardation above a crystal of moderate to high birefringence. (C) Same for crystal of low birefringence.

The number and disposition of the isochromes seen in an acute bisectrix figure depend upon (1) the crystal's thickness, t, (2) the difference between its refractive indices for the vibration directions of the slow and fast rays traveling upward along the acute bisectrix—that is, $(\beta - \alpha)$ for positive crystals and $(\gamma - \beta)$ for negative crystals, (3) the value of $2E$, and (4) the numerical aperture of the objective being used. Let Fig. 9–15B and C represent equally thick plates of two different minerals that have comparable values for $2E$ but that differ considerably in their value for $(\gamma - \beta)$, this value for the crystal in Fig. 9–15C being one-quarter what it is for the crystal in Fig. 9–15B. Consequently, if the retardation, $t(\gamma - \beta)$, equals 1400 mμ (third-order green) for the ray traveling along X in Fig. 9–15B, that for the comparable ray in Fig. 9–15C equals 350 mμ (first-order yellow); the parenthesized interference colors appear at the center of the interference figures. Note that in Fig. 9–15C the first-order red isochrome thus compares in shape and position to the fourth-order red isochrome in Fig. 9–15B. If the crystal in Fig. 9–15C were four times as thick, the surfaces of equal retardation for the emergent rays would resemble those in Fig. 9–15B. If either crystal had a smaller (or larger) value for $2E$, the spots of first-order black marking the points of emergence of rays that traveled along the optic axis while in the crystal—that is, the melatopes—would be closer (or farther) from the center of the field of view. Since the retardation colors become of higher order in directions outward from the melatopes, the shape and number of isochromes are affected by the location of the melatopes. In Fig. 9–16B the outer circle represents the field of view for an objective of N. A. 0.85; for an objective of N. A. 0.65, only the portion of the figure within the dashed circle would be visible.

The orthographic projection of the privileged directions for ray paths emerging in the field of view of an interference figure may be approximately determined by a two-dimensional application of the Biot-Fresnel rule. For example, in Fig. 9–16A dashed construction lines may be drawn from point a to either melatope; the orthographically projected vibration directions for point a are then the bisectors of the angles between the two dashed lines.

Isogyres. The shape and distribution of the areas of extinction—that is, isogyres—in an interference figure depend upon the distribution within the field of view of the rays for which the orthographically projected

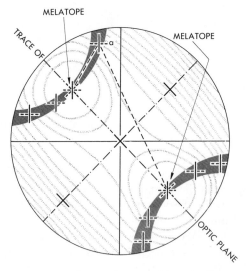

Fig. 9-16A. Centered, acute bisectrix figure at 45 degrees off extinction position. The two-dimensional analog of the Biot-Fresnel law is applied to point a to determine (approximately) the privileged directions for rays emerging there. At several points in the field of view the privileged directions for the rays emerging there are also shown.

privileged directions are N–S or nearly so. Utilizing the Biot-Fresnel rule, all such points of emergence could be located in Fig. 9–16A; their distribution would outline the shape of a hyperbola whose vertices coincide with the melatopes. Assuming a N–S vibrating polarizer, only rays vibrating N–S would have emerged at these points; consequently, the E–W privileged directions in Fig. 9–16A have been drawn dashed to indicate the absence of a light ray vibrating E–W. Thus, since the interference figure is viewed with the E–W polarizing analyzer inserted, these hyperbolic areas are sites of extinction and appear as black brushes.

If the microscope stage containing the crystal is rotated while the interference figure is being observed, the shape and location of the isogyres change drastically whereas the image of the isochromes merely rotates to the extent that the stage does, their shapes remaining constant. Fig. 9–16B

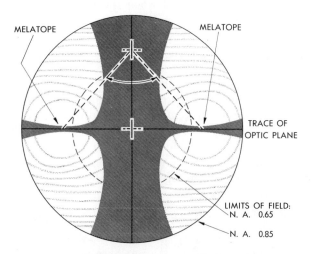

Fig. 9-16B. Fig. 9–16A as seen if the microscope stage is rotated 45 degrees counterclockwise. Extinction occurs in the areas where rays emerge that vibrate parallel to the polarizer. The dashed circle marks the limits of the field of view if an objective of N. A. 0.65 is used instead of one of N. A. 0.85.

illustrates the appearance of the interference figure in Fig. 9–16A after the microscope stage was rotated until the crystal's optic plane was E–W. The Biot-Fresnel rule can again be utilized to locate the points of emergence of rays associated with N–S and E–W privileged directions; one representative construction is shown in the figure. The isogyres are now in the shape of a cross. Generally, the thinner bar of this cross coincides with the trace of the optic plane within the interference figure; the thicker bar marks the trace of the principal plane containing the acute bisectrix and the optic normal.

For biaxial interference figures in general, whenever the crystal is rotated so that the trace of a principal plane within the field of view is parallel to a N–S or E–W cross hair, a more or less straight isogyre is developed along the length of this trace. Whenever an isogyre passes through the cross-hair intersection the crystal has been rotated into an extinction position (if viewed under an orthoscopic set-up.)

Crystal with Large 2V. Concentric with the circles that delimit the field of view in Figs. 9–16A and B, draw a light circle of one-half their radius. Assuming these two smaller circles to be the edge of the field of view (that is, the portions of the interference figure outside these circles are no longer visible), the interference figures within them represent those of a crystal whose 2E angle is so large that the optic axes do not emerge in the field of view. In spite of the fact that the melatopes do not appear within the field of view, the trace of the optic plane may be located from (1) the characteristic pattern of the isochromes or (2) the location of the thinner bar of the cross produced when the trace of the optic plane is E–W or N–S (for example, Fig. 9–16B).

Optic Axis-centered Figures

Crystals oriented with one optic axis precisely upright—that is, normal to the plane of the stage—yield interference figures in which this optic axis emerges as a melatope at the center of the field of view (Fig. 9–17). The second optic axis will emerge at various distances outward from the center of the field, depending on the value of 2E for the mineral; in the example (Fig. 9–17A), the second optic axis emerges slightly beyond the field of view. The dashed construction line in Fig. 9–17A represents the trace of the optic

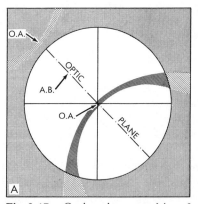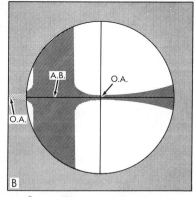

Fig. 9-17. Optic axis-centered interference figure. The crystal producing this figure is oriented with an optic axis perpendicular to the microscope stage (and with its optic plane NW-SE). (B) Appearance of this interference figure if the stage is rotated 45 degrees counterclockwise. The thin bar of the cross now formed marks the trace of the optic plane in the field of view; the thick bar marks the trace of the principal plane containing the acute bisectrix and the optic normal. The intersection of these two bars marks the point of emergence of the acute bisectrix, which, in this case, falls within the field of view.

plane. Along this trace are located the point of emergence of the acute bisectrix, which is always to the convex side of the isogyre, and the point of emergence of the obtuse bisectrix (not shown), which is always on the concave side of the isogyre.

Rotation of the crystal by turning the microscope stage causes the isogyre of Fig. 9–17A to rotate about the melatope as an axis. However, when the crystal's optic plane becomes E–W or N–S, a cross is formed whose thinner bar locates the optic plane (Fig. 9–17B). The intersection of the central lines for the thin and thick bar of this cross marks the emergence of the acute bisectrix within the field. If the acute bisectrix does not emerge within the field, no cross will form within the field.

Obtuse Bisectrix-centered Figures

Biaxial crystals broken or cut perpendicular to an obtuse bisectrix yield interference figures at the center of which the obtuse bisectrix emerges. Such figures resemble the acute bisectrix figures for minerals with large values of $2V$ (see p. 176); even the pattern of their isochromes is similar. Their differentiation will be discussed on page 179.

Optic Normal Figures

Crystals oriented on the microscope stage so that Y, their optic normal direction, stands upright—that is, perpendicular to the stage—yield optic normal figures. For such figures, if the stage is rotated until a principal plane (ZY or YX) is parallel to the privileged direction of either polarizer, the isogyres form a broad, poorly defined black cross that almost completely fills the field of view. In appearance it resembles the cross for the uniaxial flash figure (Fig. 7–21C). As is the case for the uniaxial flash figure, a very small rotation of the microscope stage causes the cross to break up into hyperbolas that rapidly leave the field of view. The distribution of the isochromes for optic normal figures is also similar to that for the uniaxial flash figure (Fig. 7–22A). Theoretically, the interference color at the edge of field of view is of lower order at the points marking the direction of the acute bisectrix than at the points marking the direction of the obtuse bisectrix. In most cases, however, this is difficult to observe.

Semirandom Figures

A semirandom plate was defined (p. 163) to be one cut parallel to one of the three principal vibration axes, X, Y, Z. Such a plate is necessarily perpendicular to the principal plane containing the remaining two principal axes. For example, the lightly shaded upper and lower surfaces of the crystal in Fig. 9–18 were cut parallel to OY and are consequently perpendicular to the dotted principal plane ZX. In the interference figure of this crystal, therefore, the trace of this principal plane passes through the cross-hair intersection. This trace (dashed in Fig. 9–18) divides the interference figure into two mirror images, a property that in part permits the trace of a principal plane within a biaxial interference figure to be recognized. For a semirandom section, the trace of a principal plane within the field of view can be more or less precisely located by rotating the microscope stage until an isogyre becomes straight and parallel to one of the cross hairs. The cross hair then coincides with the trace of this principal plane—for example, the E–W cross hair in Fig. 9–18B.

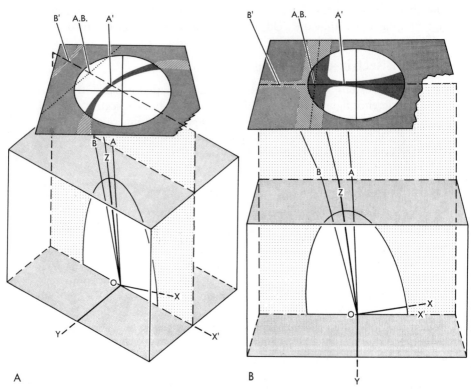

Fig. 9-18. (A) Interference figure for a grain resting on an $X'Y$ semirandom section (shaded). The light cone that passes through and emerges from the crystal has been omitted from the drawing. The light rays of the cone that travel along the optic axes OA and OB within the crystal take the paths AA' and BB' after emergence in air. The principal plane ZX (stippled) is at 45 degrees to the privileged direction of the polarizer. (B) Principal plane ZX is at 90 degrees to the privileged direction of the polarizer.

Random Figures

A principal plane cut through a biaxial crystal is perpendicular to the remaining two principal planes. A semirandom plane is perpendicular to only one of the three principal planes. The interference figure from a crystal resting upon a principal plane was therefore characterized by the traces of two principal planes passing through the cross-hair intersection (Fig. 9–16A), if it rested upon a semirandom plane, however, the trace of only one principal plane was observed to pass through the cross-hair intersection (Fig. 9–18). A random plane, in contrast, is not parallel to principal axis X, Y, or Z and is therefore not perpendicular to any of the three principal planes. Thus, in the interference figure from a crystal resting on a random plane, the trace of no principal plane is observed to pass through the cross-hair intersection.

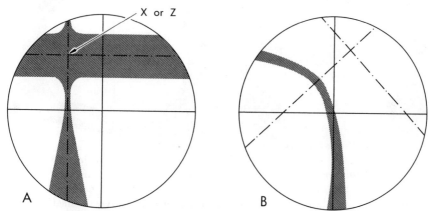

Fig. 9-19. Random figure (that is, off-centered acute bisectrix figure). (A) With traces of its principal planes (dash-dotted lines) parallel to the cross hairs. (B) Same figure with stage rotated clockwise until an isogyre passes through the cross-hair intersection.

The traces of principal planes may or may not appear in the interference figure from a grain resting on a random plane. If they do, they become the sites of more or less straight isogyres when the crystal is rotated until these traces are parallel to the cross hairs (Fig. 9–19A). For this position of the stage, if two principal traces appear in the field (as in Fig. 9–19A), the isogyres form a cross whose center marks the point of emergence of principal axis X or Z. If the stage is rotated until the isogyre passes through the cross-hair intersection (Fig. 9–19B)—at which orientation the crystal will be at an extinction position—a typical random, biaxial, interference figure may be recognized by an isogyre that follows parallel to neither cross hair.

Differentiation of Interference Figures

The different interference figures representing different orientations of a biaxial crystal may not always be distinguishable. For example, if $2V$ is large for the crystal, a centered acute bisectrix figure may be readily mistaken for a centered obtuse bisectrix figure. The latter figure may also be mistaken for an optic normal figure. Distinction between these figures may sometimes be made by determining δ, the angle of rotation of the stage necessary to cause the isogyres to move from an extinction position (Fig. 9–20A) until the center line through the hyperbolic shadow (dashed in Fig. 9–20B) is tangent to the edge of the field of view. Let δ_{NW}, δ_{SW},

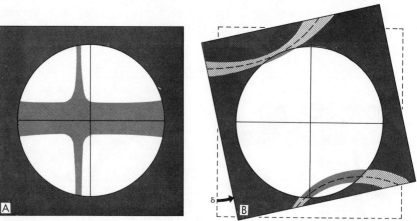

Fig. 9-20A-B. (A) A slightly off-centered biaxial interference figure in an extinction position. For this position an isogyre passes through the cross-hair intersection. (B) Appearance of this same figure after the stage has been rotated counterclockwise through δ, an angle just sufficient to cause the center of an isogyre (dashed line) to be tangent to the NW edge of the field of view.

δ_{SE}, and δ_{NE} denote the angles of stage rotation necessary to produce such tangency to the edges of the NW, SW, SE, and NE quadrants of the field of view, respectively. For example, to alter the interference figure from Fig. 9–20A to 9–20B required a counterclockwise rotation through the angle δ_{NW}. Note that a counterclockwise rotation through an angle slightly larger than δ_{NW} would have caused the central line of the lower hyperbolic isogyre (dashed in Fig. 9–20B) to become tangential to the edge of the SE quadrant; thus δ_{SE} is larger than δ_{NW}. Similarly, clockwise rotations from the extinction position of Fig. 9–20A would permit measurement of δ_{SW} and δ_{NE}. These four angles—δ_{SE}, δ_{NW}, δ_{SW}, and δ_{NE}—are equal only for perfectly centered interference figures; that is, those for which a principal axis emerges at the cross-hair intersection. Even for slightly off-centered figures, however, δ— the average of these four angles—permits V to be determined by use of the curves (Fig. 9–20C) developed by Kamb (1958), where V represents the angle (within the crystal) between the centered (or nearly centered) principal axis and the optic axis. In order to decide which curve in Fig. 9–20C to use, it is necessary to know the approximate value of $β$ for the unknown and to observe the interference figure with an objective of numerical aperture of 0.85. The principal axis that emerges at the center of the field of view is then (1) the acute bisectrix, if V is less than 45 degrees; or (2) the obtuse bisectrix, if V exceeds 45 degrees; or (3) the optic normal, if δ is so small that a horizontal line through its value falls below the curve. The accuracy to which $2V$ can be determined by this method depends in part upon the degree to which the observed interference figure is truly centered.

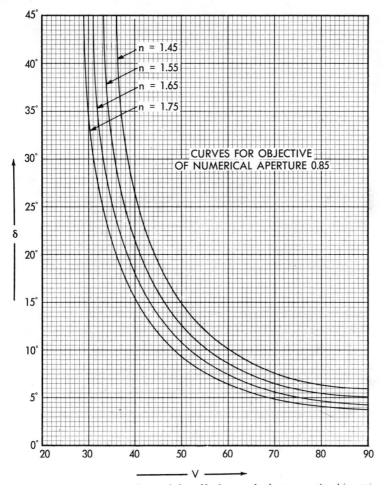

Fig. 9-20C. Chart for determining V, the angle between the bisectrix emerging at the center of the field of view and an optic axis, provided the value of δ has been measured from a centered interference figure viewed with an objective of numerical aperture 0.85. (After Kamb, 1958.)

Some biaxial interference figures may be mistaken for uniaxial figures, but others may be definitely distinguished. Biaxial figures for which an optic axis occurs within the field of view are generally easy to differentiate from uniaxial figures. Centered acute bisectrix figures of minerals with very small $2V$ (1 to 2 degrees), however, are difficult to differentiate from a uniaxial figure, for the separation of the melatopes that occurs when the optic plane is in the 45 degree position is observed only with difficulty. This separation may be better observed, provided the grain is large enough, by substituting a medium-power objective (10x, N. A. 0.25) for the high-power objective and thus obtaining an enlarged view of the center of the interference figure.

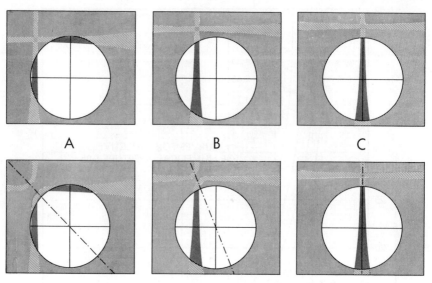

Fig. 9-21. Similarity between an off-centered, uniaxial figure (upper diagrams) and a biaxial figure with small $2V$ (lower diagrams) for the crystal (A) at 45 degrees off extinction, (B) at 22 degrees off extinction, and (C) at extinction.

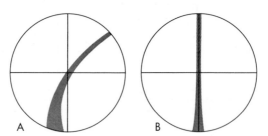

Fig. 9-22. Comparison of the shape and alignment of two isogyres after the stage has been rotated until each passes through the cross-hair intersection. (A) An unequivocally biaxial isogyre is characterized by a curved shape and nonparallelism to either cross hair. (B) An isogyre that may represent either a uniaxial crystal or a semirandom section of a biaxial crystal. Note its longitudinal bisection by the N-S cross hair.

When the optic axis does not occur within the field of view, a biaxial crystal with small $2V$ may be indistinguishable from a uniaxial crystal (Fig. 9–21). However, some interference figures, even though no optic axis falls within the field, may be clearly differentiated by rotating the stage until an isogyre passes through a cross-hair intersection. For this position, an undoubtedly biaxial crystal will be distinguished by an isogyre that does not parallel either cross hair (Fig. 9–22A). In contrast, an isogyre that is bisected along its length by a cross hair (Fig. 9–22B) may indicate either a uniaxial crystal or a biaxial crystal resting on a semirandom section.

Relationship to Orthoscopic Observations

From the interference figure of a crystal plate, the microscopist can deduce several features that the crystal would exhibit if viewed orthoscopically. For such deductions, the key area to observe is the intersection of the

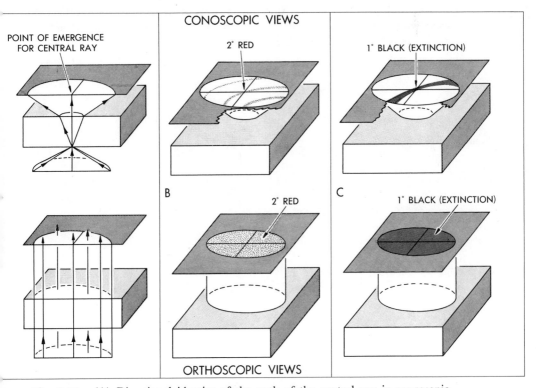

Fig. 9-23. (A) Directional identity of the path of the central ray in conoscopic illumination (upper) with all the ray paths in orthoscopic illumination (lower). (B) Identification of the interference color in the isochrome passing through the cross hairs in the interference figure (upper) with the interference color exhibited by the entire plate when viewed orthoscopically (lower). (C) Passage of an isogyre through the cross-hair intersection (upper) signifies that the entire crystal is at extinction when viewed orthoscopically (lower).

cross hairs—that is, where the central ray of the cone of light that passed through the crystal emerges in the field of view. This central ray essentially passes through the crystal in the same direction as do *all* the rays when the crystal is viewed orthoscopically (Fig. 9–23A). Consequently, the retardation color represented by that isochrome of the interference figure which passes through the cross-hair intersection is also the retardation color of the crystal plate (feathered edges excepted), if viewed orthoscopically (Fig. 9–23B). Similarly, the crystal is at a position of orthoscopic extinction whenever an isogyre passes through the cross-hair intersection (Fig. 9–23C).

DISPERSION AND CRYSTALLOGRAPHIC ORIENTATION OF X, Y, AND Z

Dispersion of the Optic Axes

Biaxial minerals possess different values of α, β, and γ for different wavelengths of light. Thus for C, D, and F light, the principal refractive indices may be symbolized α_C, α_D, α_F; β_C, β_D, β_F; and γ_C, γ_D, γ_F. In detailed

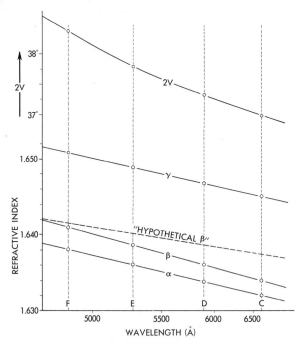

Fig. 9-24. Change in value of α, β, γ and therefore of $2V$ for different wavelengths of light in the mineral barite. (Based on data from Winchell, 1931.) If β had changed as indicated by the dashed line labeled "hypothetical β," the slope of the $2V$ curve would have been reversed and the angle $2V_C$ would then exceed $2V_F$.

optical measurements, all nine may be determined. If these indices, or indices for additional wavelengths, are plotted on Hartmann dispersion paper, straight lines usually result (Fig. 9–24). For barite, as Fig. 9–24 illustrates, β more closely approaches α in value for the longer wavelengths of light. As a consequence, $2V$ for barite decreases for the longer wavelengths. Thus $2V_r$ is less than $2V_v$, where the subscripts r and v respectively denote red and violet light (the opposite ends of the visible spectrum). The foregoing inequality is generally abbreviated to read $r < v$. On the other hand, if β had diverged from α for the longer wavelengths (dashed hypothetical line), $2V$ would have increased for the longer wavelengths; $r > v$ describes such a situation. In biaxial minerals, therefore, the optic axes do not maintain a constant position for all wavelengths of light. The degree of their dispersion is expressed by the slope of the $2V$ curve in Fig. 9–24. Depending upon the extent by which r exceeds v (or v exceeds r), the dispersion of the optic axes is classified as weak, moderate, or strong. Of course, if r equals v, there is no dispersion at all. All crystals of barite exhibit weak dispersion, with $r < v$; all crystals of sillimanite exhibit strong dispersion, with $r > v$. The degree of dispersion thus helps to identify a mineral.

Orthorhombic Crystals

In the orthorhombic system the three, mutually perpendicular, crystallographic axes a, b, and c coincide with the three, mutually perpendicular, principal vibration directions X, Y, and Z (not necessarily respectively) for all wavelengths of light. For example, in all barite crystals $a = Z$, $b = Y$, and $c = X$ whereas in all olivine crystals $a = Z$, $b = X$, and $c = Y$. In orthorhombic crystals, therefore, only dispersion of the optic axes occurs.

Normal Orthorhombic Dispersion. This dispersion of the optic axes in orthorhombic crystals is best observed from centered or near-centered acute bisectrix figures. In such figures for barite (Fig. 9–25A) and sillimanite (Fig. 9–25B) this dispersion causes the melatopes, A_r and A_v, and the isogyres for red and violet light to be displaced from each other. Sillimanite, since $r \gg v$, shows the greater displacement. The melatopes and isogyres for intermediate wavelengths are located between those for red and violet. Areas of overlap between the red and violet isogyres mark areas that are isogyres for all wavelengths—that is, areas of extinction for light of all wavelengths. For areas where only the isogyre for red light (shaded in Fig. 9–25) occurs, red light is extinguished at the analyzer whereas violet or blue light is transmitted. For areas where only the isogyre for violet light (ruled in Fig. 9–25) occurs, violet or blue light is extinguished at the analyzer whereas red is transmitted. Consequently, within interference figures of crystals exhibiting at least a weak dispersion of the optic axes, the black isogyre is fringed—especially in the region of the melatope—by a reddish tint on one side and a violet (or blue) tint on the other. The red fringe marks the melatope and a portion of the isogyre for violet light;

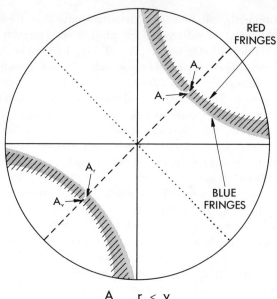

A   r < v

Fig. 9-25. Origin of dispersion fringes in acute bisectrix figures (A) for barite and (B) for sillimanite. The isogyre locations for red light are shown as shaded areas; for blue light, as ruled areas. Red light emerging along the red (shaded) isogyre is extinguished at the analyzer whereas blue light is transmitted. Blue light emerging along the blue (ruled) isogyre is extinguished whereas red light is transmitted. Where both isogyres are superimposed, all wavelengths are extinguished; these areas thus appear black. Where they are not superimposed, a blue coloration is developed along the red isogyre and melatope whereas a red coloration is developed along the blue isogyre and melatope. The dotted and the dashed lines represent the traces of principal planes.

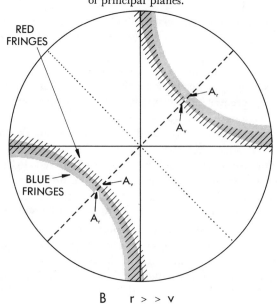

B   r > > v

the violet (or blue) fringe marks the melatope and a portion of the isogyre for red light. These color fringes are, however, easily overlooked.

The dispersional color fringes for orthorhombic crystals are always symmetrically disposed with respect to the trace of the optic plane (dashed in Fig. 9–25) and the principal plane normal to it (dotted in Fig. 9–25). If sufficiently distinct, this symmetry of the dispersion colors with respect to the traces of *two* principal planes permits the crystal system of a biaxial crystal to be unequivocally established as orthorhombic.

Crossed Axial Plane Dispersion. The orthorhombic mineral brookite displays an extraordinary dispersion of the optic axes, since its refractive index for vibrations parallel to the *c* axis is the lesser index (α) for wavelengths less than 5550 Å but the intermediate index (β) for wavelengths longer (Fig. 9–26). Consequently, for the shorter wavelengths the optic plane

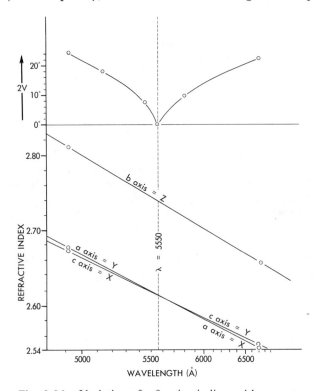

Fig. 9-26. Variation of refractive indices with respect to wavelength for light vibrating parallel to the *b* axis, *a* axis, and *c* axis in brookite. Top scale : variation of the optic angle 2*V* for different wavelengths of light.

(stippled in Fig. 9–27A) is parallel to the *b* and *c* axes—that is, to the (100) crystal plane—whereas for wavelengths that are longer than 5550 Å the optic plane (shaded) is parallel to the *a* and *b* axes—that is, to the (001) plane. For the 5550 Å wavelength the crystal is uniaxial $(+)$, the *b* axis being the optic axis. Within the stippled (100) plane, the optic angle is smaller for the longer wavelengths of light; thus, $r < v$. Within the (001) plane, the optic angle is larger for the longer wavelengths (*cf.* $\lambda = 6500$ Å and $\lambda = 6000$ Å); thus $r > v$.

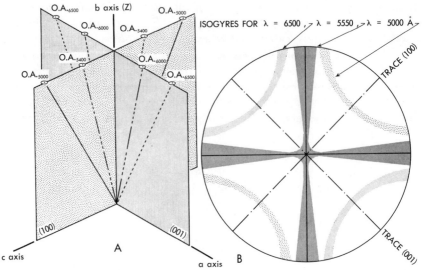

Fig. 9-27. Crossed axial plane dispersion in brookite. (A) Location of the optic axes $(O.A._{\lambda})$ for four selected wavelenghts of light. The optic plane for wavelengths shorter than 5550 Å is stippled; for wavelengths longer than 5550 Å it is shaded. For light of wavelength 5550 Å, $2V$ equals $0°$, the single optic axis coinciding with the crystallographic b axis. (B) Location in the acute bisectrix figure of the isogyres for 5000 Å, 5550 Å, and 6500 Å only.

The acute bisectrix figure of brookite (b axis perpendicular to plane of stage) consequently shows a type of dispersion known as *crossed axial plane dispersion*. Illuminated by white light, two, mutually perpendicular, optic planes (Fig. 9–27B) are developed, one for the wavelengths of light shorter than 5550 Å, the other for the wavelengths longer. The resultant interference figure is complicated and typical of crossed axial plane dispersion. Illuminated with monochromatic light, the interference figures appear normal, being uniaxial $(+)$ if the wavelength is 5550 Å but biaxial $(+)$ for longer or shorter wavelengths. The isogyres for only a few wavelengths of light are shown in Fig. 9–27B.

Monoclinic Crystals

In monoclinic crystals only one principal axis, X, Y, or Z, coincides (for all wavelengths of light) with the symmetry dictated b axis*. The

---

* By definition, in monoclinic crystals the b axis coincides with the single twofold axis or with the normal to the single plane of symmetry, or with both.

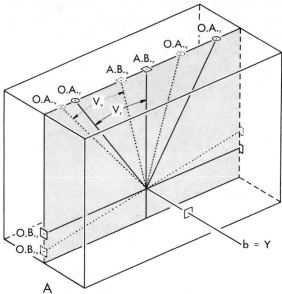

A

Fig. 9-28. Inclined dispersion in a monoclinic crystal for which $b = Y$ and $V_r > V_v$—that is, $r > v$. (A) Dispersion of the optic axes, acute bisectrix, and obtuse bisectrix. Their positions for red light and violet light are indicated by subscripts. (B) Acute bisectrix-centered interference figure of the crystal, optic plane at 45-degree angle to vibration of polarizer. The blue-tinted fringes (shaded) mark the locations of the red isogyres. The red-tinted fringes (ruled) mark the edges of the violet isogyres.

(Degree of dispersion has been highly exaggerated.)

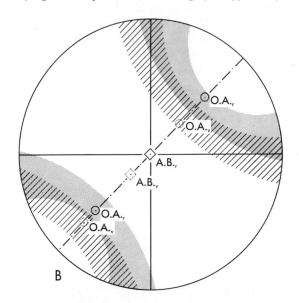

B

remaining two principal axes always lie within a plane perpendicular to the *b* axis but occupy different locations within this plane for different wavelengths (Fig. 9–28A). They may coincide with the *a* or *c* axis only by chance. The dispersion of these two principal axes (coupled with a dispersion of the optic axes similar to that discussed for orthorhombic crystals; see page 184) reduces the symmetry of the distribution of the dispersional fringes in the interference figures of monoclinic crystals. In some cases the distribution of these dispersional fringes is sufficiently distinct to permit not only a differentiation of a monoclinic crystal from an orthorhombic crystal but also, if the crystal is monoclinic, determination of whether the *b* axis coincides with (1) Y, the optic normal, (2) the obtuse bisectrix, or (3) the acute bisectrix.

Inclined Dispersion (b = Y). If the Y principal axis of the indicatrix coincides with the symmetry controlled *b* axis, this axis is the Y direction for light of all wavelengths; that is, there is no dispersion of Y (Fig. 9–28A). Moreover, for all wavelengths the optic plane is normal to this axis, although, within this plane, the location of the acute and obtuse bisectrices varies for the different wavelengths of light. In consequence, this dispersion of the bisectrices (in addition to the dispersion of the optic axes) reduces the symmetry of the disposition of the dispersion fringes from what it was in the orthorhombic case. As shown in Fig. 9–28B, the dispersion colors are bilaterally symmetrical with respect to the trace of only one plane (in contrast to orthorhombic crystals, which were symmetrical with respect to two). This trace, dot-dashed in Fig. 9–28B, represents the trace of the optic plane for all wavelengths. This type of dispersion, called inclined dispersion, is thus characterized by a common optic plane for all light but different acute bisectrix and melatope positions for the different wavelengths.

Parallel or Horizontal Dispersion (b = obtuse bisectrix). If the obtuse bisectrix (O.B.) coincides with the *b* axis, this axis serves as the obtuse bisectrix for light of all wavelengths. On the other hand, the optic normal (Y) and the acute bisectrix occupy different positions within the plane normal to the *b* axis for the different wavelengths; Fig. 9–29A illustrates their positions for violet and red light only. The optic planes for the different wavelengths (shaded lightly for red light and more heavily for violet or blue light in Fig. 9–29A) intersect at a common hinge, the *b* axis. Within an acute bisectrix-centered figure of such a crystal (Fig. 9–29B), these optic planes intersect the field of view as a series of parallel traces; hence the term parallel or horizontal dispersion. Within such an interference figure the parallelism of these traces of the optic planes can be deduced from the distribution of the dispersional color fringes; for example, the line (dotted in Fig. 9–29B) joining the blue-tinted fringes (which locate the melatopes for red light) and that (dashed) line joining the red-tinted fringes (which locate the melatopes for violet light) are parallel. Note that the color fringes are bilaterally symmetrical to only one line (dot-dashed in Fig. 9–29B); it represents the trace of the plane perpendicular to the *b* axis.

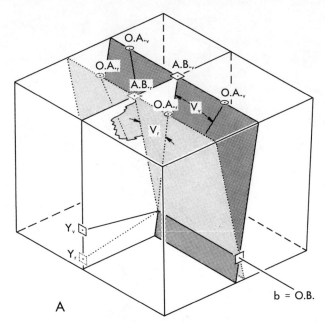

Fig. 9-29. Parallel or horizontal dispersion in a monoclinic crystal for which $b = $ O.B. and $r < v$. (A) Crystal in orientation so as to yield an acute bisectrix-centered figure. The positions of the acute bisectrix (A.B.) and optic normal (Y) are shown for the opposite ends of the visible spectrum—red and violet light, respectively. The optic plane for red light (lightly shaded) and violet light (heavily shaded) intersect at the b axis, which is the obtuse bisectrix for all wavelengths. (B) Acute bisectrix-centered figure of this crystal (at 45 degrees off extinction position). The blue-tinted fringes (shaded) mark the locations of the red melatopes; the red-tinted fringes (ruled) mark the locations of the violet melatopes. The trace of the optic plane for red light is shown as a dotted line; that for violet light, as a dashed line.

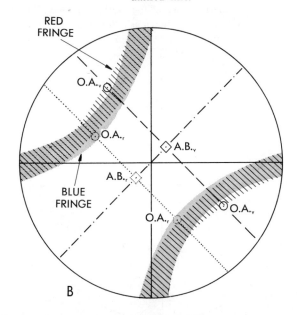

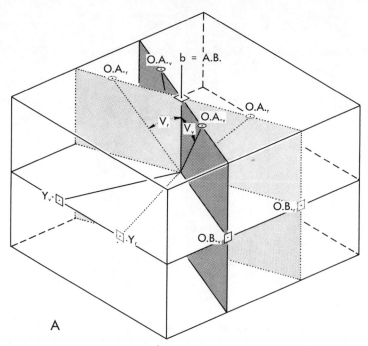

A

Fig. 9-30. Crossed dispersion in a monoclinic crystal for which $b = $ A.B. and $r > v$. (A) Crystal in orientation so as to yield an acute bisectrix-centered figure. The locations of the obtuse bisectrix (O.B.) and optic normal (Y) are shown for red and violet light, respectively. The optic plane for red light (lightly shaded) and for violet light (heavily shaded) intersect at the b axis, which is the acute bisectrix for all wavelengths. (B) The acute bisectrix-centered figure of this crystal (at 45 degrees off extinction). The blue-tinted fringes (shaded) mark the locations of the red melatopes; the red-tinted fringes (ruled) mark the locations of the violet melatopes. The trace of the optic plane for red light is shown as a dotted line; that for violet light, as a dashed line.

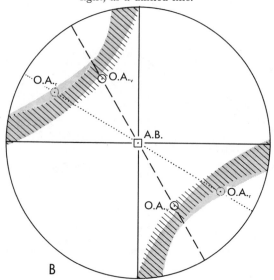

B

Crossed Dispersion (b = acute bisectrix). If the acute bisectrix (A.B.) coincides with the *b* axis, this axis serves as the acute bisectrix for light of all wavelengths; the optic normal and obtuse bisectrix, however, occupy different positions (within the plane normal to this *b* axis) for the different wavelengths. Fig. 9–30A illustrates their positions for violet and red light only. For other wavelengths their positions would be intermediate. As may be seen, the optic planes for the different wavelengths intersect in a common axis, this axis representing the acute bisectrix for all wavelengths. Consequently, as viewed in the interference figure (Fig. 9–30B), the traces of the optic planes for the different wavelengths cross each other at the point of emergence of their common acute bisectrix; hence the term crossed dispersion. The disposition of the dispersion fringes in Fig. 9–30B reveals no symmetry plane to be normal to the field of view. The normal to the field of view, however, can be regarded as an axis of two-fold symmetry. This normal coincides with the *b* axis of the crystal.

Differentiation of Crystal Systems

The symmetry of the dispersional color fringes in an acute bisectrix-centered interference figure is very difficult to observe, rarely being as clear as in the foregoing illustrations. Under favorable conditions, however, these dispersion effects permit the crystal system to be identified as ortho-rhombic, monoclinic, or triclinic, as the case may be. In general, the color fringes will be symmetrical with respect to *two*, mutually perpendicular planes (Fig. 9–31A) if the crystal is orthorhombic. They will be sym-metrical to only *one* such plane (Fig. 9–31B) or to a two-fold axis (Fig. 9–31C) if the crystal is monoclinic. Moreover, in monoclinic crystals the *b* axis can be located, since it coincides with the normal to this plane of symmetry or with the two-fold axis, whichever was found; one could therefore deduce

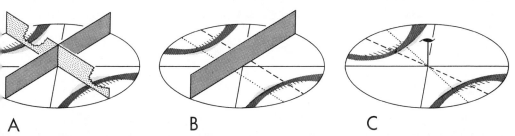

A   B   C

Fig. 9-31. Differentiation of orthorhombic and monoclinic crystals from the symmetry of the dispersional color fringes in their acute bisectrix-centered figures. Orthorhombic crystals can always be considered to have two planes of symmetry perpendicular to the interference figure (A). Monoclinic interference figures are symmetrical only with respect to one such plane (B) or to a twofold axis perpen-dicular to the plane of the interference figure (C). The *b* axis of monoclinic crystals either is the direction perpendicular to this plane (B) or it coincides with the two-fold axis (C).

from Figs. 9–28B, 9–29B, and 9–30B, even if they were unlabeled, that $b = Y$, $b = $ O. B., and $b = $ A. B., respectively.

The dispersional color fringes in the acute bisectrix-centered figures of triclinic crystals are asymmetric, lacking either bilateral symmetry or a two-fold axis normal to the plane of the interference figure. In addition to the dispersion of the optic axes, all three principal vibration axes, X, Y, and Z, undergo dispersion. As may be deduced, the three crystallographic axes, a, b, and c, in triclinic crystals do not coincide with the X, Y, and Z axes (except in a chance instance for a particular wavelength of light).

Abnormal Interference Colors

The abnormal interference colors seen in biaxial crystals at 45 degrees off extinction are similar in origin to those previously discussed for uniaxial crystals (p. 149). However, abnormal interference colors of entirely different origin from these may be seen in biaxial crystals that are within a degree or two of an extinction position. The phenomenon is observed chiefly in biaxial crystals for which the dispersion of the principal vibration axes (X, Y, and Z) or of the optic axes, or of both, is so pronounced that the privileged directions for normal incidence on a given face are not exactly the same for all wavelengths of light. Close to extinction positions, therefore, anomalous interference colors may be observed instead of blackness, the crystal never being simultaneously extinct for all the wavelengths in white light.

| TEN | O PTICAL EXAMINATION |
| OF BIAXIAL CRYSTALS |

In all respects sample preparation of biaxial crystals is the same as for uniaxial crystals (see p. 123). The crushed grains, if no dominant cleavage exists, will rest on fracture surfaces that are variously oriented with respect to principal axes X, Y, and Z and the two optic axes. Whenever such a surface is parallel to a principal axis, this axis will serve as a privileged direction for the grain. Consequently, the grain will exhibit a principal refractive index (α, β, or γ) whenever this principal axis (X, Y, or Z, respectively) is made parallel to the privileged direction of the polarizer. Table 10-1 summarizes the refractive indices measurable from grains resting on the different types of surfaces previously discussed (see p. 163 and Table 9–2). If, as for the micas, a dominant cleavage causes all grains to rest on optically identical surfaces, ground glass may be liberally added to the oil mount before adding the mineral. Mineral grains propped up by grains of glass will then diversify the optic orientations. The spindle stage (Rosenfeld, 1950; Wilcox, 1959) also solves this problem very nicely and inexpensively.

Barring a dominant cleavage, the majority of grains in a powder mount are likely to rest on random planes. Measurement of their indices is of little value since the measured values, α' and γ', will differ from one grain to the other. The significant indices α, β, and γ, on the other hand, can only be measured from the small minority of grains resting on principal or semi-random planes (Table 10–1). The microscopist thus faces the task of searching out these grains from among the myriads of random orientations in the mount. Techniques for recognizing the grain types and subtypes will be discussed incidental to the problem of measuring α, β, and γ. First, however, the determination of the optic sign of biaxial minerals will be discussed, since, in practical optical mineralogy, this is usually one of the first optical characteristics to be established.

DETERMINATION OF BIAXIALITY

Using crossed nicols and a low-power objective, search for the largest grain exhibiting a first-order gray (or lesser) retardation color at 45 degrees off extinction. This grain is most likely to rest on a YC plane since $\varDelta_{yc}$ (Table 10–1) is less than the retardation of any other grain type; thus

Table 10–1

RELATIONSHIP OF THE ORIENTATION OF A GRAIN'S PLANE OF REST
TO ITS REFRACTIVE INDICES AND RETARDATION

Plane of rest	Refractive indices measurable for privileged directions of grain	Retardation for a grain of thickness t
	(1) Principal planes	
XY	α and β	$\varDelta_{xy} = (\beta - \alpha)t$
XZ	α and γ	$\varDelta_{xz} = (\gamma - \alpha)t$
YZ	β and γ	$\varDelta_{yz} = (\gamma - \beta)t$
	(2) Semirandom planes	
XZ'	α and γ'	$\varDelta_{xz'} = (\gamma' - \alpha)t$
YX'	β and α'	$\varDelta_{yx'} = (\beta - \alpha')t$
YZ'	β and γ'	$\varDelta_{yz'} = (\gamma' - \beta)t$
YC	β only	$\varDelta_{yc} = 0$
ZX'	γ and α'	$\varDelta_{zx'} = (\gamma - \alpha')t$
	(3) Random planes	
$X'Z'$	α' and γ'	$\varDelta_{x'z'} = (\gamma' - \alpha')t$

conversion from orthoscope to conoscope will usually produce a centered or near-centered optic axis interference figure from which the crystal's biaxial nature and optic sign are readily determinable. Grains less well oriented may also permit these determinations; in general, however, the presence of an optic axis in the field of view simplifies sign determination and, in addition, permits the biaxial nature of the crystal to be more surely recognized. In this respect, note that an off-centered figure of a biaxial crystal of small $2V$ is indistinguishable from that of an off-centered uniaxial crystal (see Fig. 9–21). Biaxiality can often be established even if no optic axes appear in the field. To test such figures, rotate the stage until an isogyre passes through the cross-hair intersection. For this position of the stage, if the isogyre is not parallel to a cross hair (as, for example, in Fig. 9–22A), the mineral is definitely biaxial. If the isogyre is parallel, it may be either biaxial or uniaxial (Fig. 9–22B).

DETERMINATION OF OPTIC SIGN

The vibration directions of rays emerging in an acute bisectrix interference figure are shown for a (−) crystal (Fig. 10–1A) and for a (+) crystal (Fig. 10–1B). Of a ray pair emerging at the same point, the vibration direction for the slower ray, since it corresponds to a greater index of refraction, is drawn longer than that for the faster ray. Insertion of an accessory plate with its N direction perpendicular to the optic plane of these crystals produces subtraction in the shaded portions of the interference figures and addition in the unshaded portions. Note that the area near the site of emergence of the acute bisectrix undergoes subtraction for negative crystals (Fig. 10–1A) but addition for positive crystals (Fig. 10–1B).

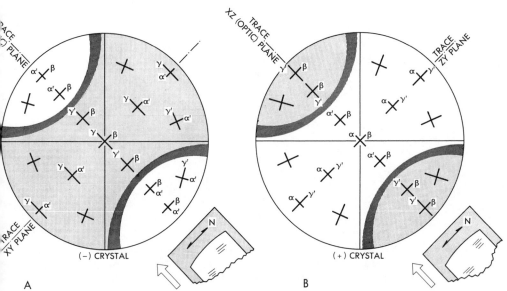

Fig. 10-1. The N and n vibration directions of the slow and fast rays, respectively, are shown at their points of emergence in the field of view as two mutually perpendicular lines, the longer representing the N direction. Shaded areas represent areas where subtraction occurs after insertion of an accessory plate with its N direction perpendicular to the optic plane. (A) Acute bisectrix-centered figure of a (−) crystal. (B) Acute bisectrix-centered figure of a (+) crystal.

The isogyres in an acute bisectrix figure at 45 degrees off extinction do not divide it into four distinct quadrants in the manner that the uniaxial cross does a uniaxial figure. Instead, only two quadrants are set apart by the isogyres, these being the unshaded quadrants in Fig. 10–1A or the shaded quadrants in Fig. 10–1B. The remaining two quadrants are now, so to speak, merged together and, in both illustrations, form an area that trends NE–SW. As for uniaxial crystals (p. 128), a line connecting the quadrants of subtraction will be parallel to the N direction of the inserted accessory if the crystal is optically negative, but perpendicular to it—thus to form an imaginary plus sign—if the crystal is optically positive. In Fig. 10–1A, for example, note that subtraction occurred in the merged quadrants and that the line joining the centers of these quadrants would be NE–SW— that is, parallel to the N direction of the inserted accessory. In Fig. 10–1B the line joining the two quadrants of subtraction would trend NW–SE and be a direction perpendicular to the N direction of the accessory.

In practical work, a crystal's optic sign is determined by locating the areas of subtraction and addition in its interference figure (1) by inserting a first-order red (gypsum) plate, if the figure contains few isochromes, or (2) by inserting the quartz wedge, if the figure possesses numerous isochromes. After its insertion, the first-order red plate causes the areas of 1° white in the region of the melatopes to become 1° yellow (Fig. 10–2, ruled areas)

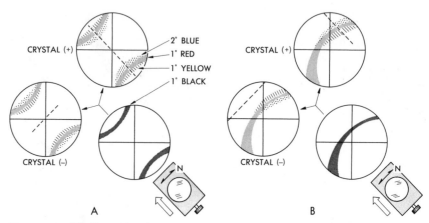

Fig. 10-2. (A) Determination of the optic sign of a centered acute bisectrix figure by insertion of a first-order red (gypsum) plate. If the optic plane trends perpendicular to the N direction of the plate, first-order yellow (ruled areas) develops on the convex side of the isogyre for negative crystals and on the concave side for positive crystals. A dashed line joining the centers of the quadrants in which subtraction occurs has been added to illustrate an alternative convention useful in sign determination. This dashed line will always be parallel to the N direction of the accessory plate for negative crystals and perpendicular to it for positive crystals. (B) Determination of the optic sign for off-centered figures. The relationship of these figures to the centered figures in (A) is easily visualized. The student may improve parts (A) and (B) by coloring the ruled areas yellow and the stippled areas blue.

along the edge of the isogyres nearest the quadrants in which subtraction has occurred, and 2° blue (stippled areas) along the edges nearest the quadrants in which addition has occurred. Viewed while the quartz wedge is being inserted, thin edge first, quadrants of subtraction (Fig. 10–3) are marked by lower order isochromes moving toward and displacing higher order ones; in the quadrants of addition this is reversed. Thus in Fig. 10–3A

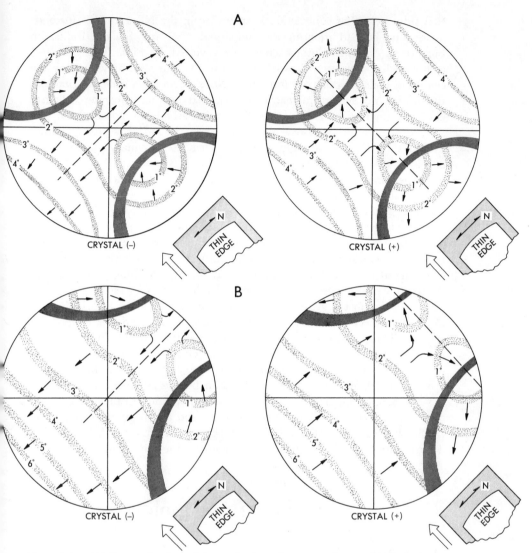

Fig. 10-3. (A) Determination of the optic sign of a centered acute bisectrix figure by insertion of a quartz wedge, thin end first. If the optic plane is perpendicular to the N direction of the wedge, subtraction occurs at the center of the field for a negative crystal (left side); in this area note that the lower order isochromes move toward the higher order isochromes, their direction of movement during insertion of the wedge being shown by the small arrows. For a positive crystal (right side), addition occurs at the center of the field, the higher order isochromes moving toward the lower ones in this region. As in Fig. 10-2, a dashed line has been drawn between the centers of the quadrants in which subtraction occurs. (B) Determination of optic sign in off-centered figures is possible if their relationship to the centered figures in (A) can be deduced. In both these off-centered figures it is necessary to determine the direction of the dashed line that joins the centers of the quadrants of subtraction in order to determine the optic sign.

(left side) note that subtraction is occurring in the NE and SW (merged) quadrants, but addition in the individual NW and SE quadrants. In Fig. 10–3A (right side) these effects are reversed. In Figs. 10–2 and 10–3 a dashed line has been drawn—for illustrative purposes only—to link the centers of the quadrants of subtraction. Note that, with a little imagination, the direction of such a line can be determined even for off-centered figures like those in Fig. 10–2B and 10–3B. If this dashed line is parallel to the N direction of the accessory plate, the crystal is biaxial negative whereas if these directions are at right angles, to form an imaginary plus sign, the crystal is biaxial positive.

The foregoing conventions are particularly useful since they require no special position for the optic plane in the field of view. Thus the stage can be turned to whatever position facilitates the observations. An alternative method of sign determination requires rotation of the stage until the crystal's optic plane is more or less perpendicular to the N direction of the accessory. For this orientation (Fig. 10–1), with respect to the interfering ray pair emerging at the site of the acute bisectrix, the ray vibrating parallel to the N direction of the accessory corresponds to β in index. The second, vibrating perpendicular to the N direction, corresponds to the index associable with the obtuse bisectrix—that is, to γ for negative crystals and to α for positive crystals. Thus, in the region of the acute bisectrix, subtraction occurs for negative crystals but addition for positive crystals when an accessory is inserted with its N direction crosswise to the crystal's optic plane.

For interference figures in which the melatopes fall far outside the field of view, Kamb's method (see p. 180) frequently permits distinction between an acute bisectrix figure with very large $2V$ and an obtuse bisectrix figure. In the latter case results will be the opposite from those for optic sign determinations on the acute bisectrix figure of the same mineral.

MEASUREMENT OF INDICES

Determination of β

The same large grain with a first-order gray retardation color that was used for the sign determination easily permits measurement of the refractive index β. If the grain rests on a surface broken parallel to a circular (γC) section of the indicatrix, a precisely centered optic axis figure (Fig. 9–17A) will be observed and the crystal will exhibit the refractive index β for all positions of the microscope stage. Thus, for such grains, it is merely necessary to convert the microscope from conoscope to orthoscope and then compare the refractive index of the grain (that is, β) with the refractive index of the oil (by Becke line or oblique illumination). The refractive index β can also be measured from grains oriented like plate A or B in Fig. 10–4, as will be discussed further on the following pages.

Determination of α and γ

With the orthoscope, search the slide (low-power objective suggested) for a relatively small grain possessing a higher order retardation color than any other. The privileged directions of such a grain are then likely to correspond to α and γ; for, if the grains of the mount had previously been sieved to produce a more or less constant thickness, this grain would (of all those in the mount) show maximum retardation colors by virtue of its birefringence $(\gamma - \alpha)$ exceeding that for grains in any other orientation. Correctly oriented grains should, under the conoscope, yield optic normal figures, since, for this orientation, X and Z should lie parallel to the plane of rest whereas Y, the optic normal, is perpendicular to it and emerges at the center of the interference figure. It is well to test the interference figure by Kamb's method to be sure it represents an optic normal figure rather than an obtuse bisectrix figure. The refractive indices, α and γ, can also be measured from grains other than those yielding optic normal figures. The recognition and correct orientation of such grains by means of their interference figures will be discussed next.

Orientation of Grains by Interference Figures

If a mineral possesses a dominant cleavage, grains resting on principal planes or on a circular section may be rare or absent. Fortunately, α, β, or γ can also be measured from grains for which at least one principal vibration direction $(X, Y, \text{ or } Z)$ is parallel to the plane of the stage. Such grains will always have a principal plane of the indicatrix perpendicular to this vibration direction, and therefore to the stage. For example, considering crystals B and C in Fig. 10–4, each has a principal vibration direction (Y for B and OB for C) parallel to the stage and therefore a principal plane perpendicular to it. For crystal A both Y and OB are parallel to the stage and therefore two principal planes are perpendicular to it.

To determine whether a grain rests with a principal plane perpendicular to the plane of the stage (and therefore with X, Y, or Z parallel to the stage), observe the grain's interference figure during rotation of the stage. Stop the rotation if an isogyre becomes straight and parallel to a cross hair, for it then marks the trace of a principal plane within the field of view (see p. 175). If this "cross-hair-parallel" isogyre passes through the cross-hair intersection, the principal plane whose trace it represents is perpendicular to the stage. If the center line of this "cross-hair-parallel" isogyre coincides with the E–W cross hair, the grain is oriented with a principal vibration direction N–S and parallel to the stage, and the index for this vibration direction is then measurable (if a N–S polarizing microscope is being used).

With appropriate rotation of the stage (as indicated by the curved arrows), the crystals in Fig. 10–4 develop interference figures like those shown at the right. A 45-degree clockwise rotation of crystal A would make its obtuse bisectrix N–S and produce an interference figure like that labeled

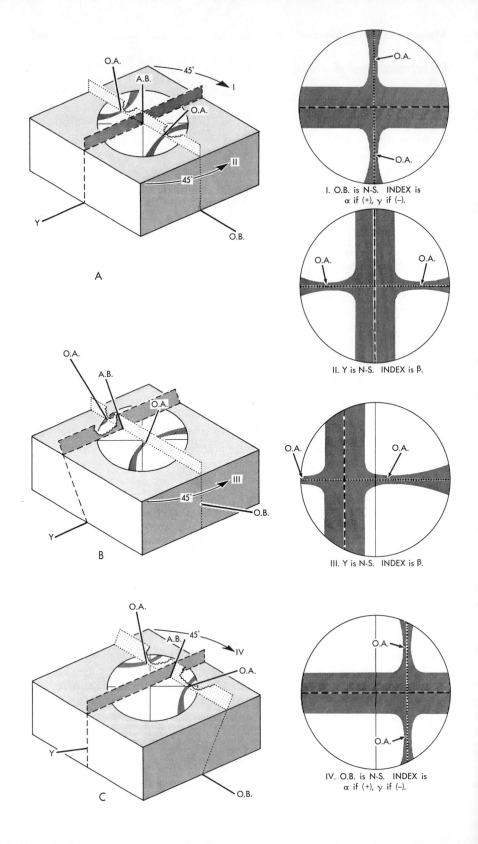

I. O.B. is N-S. INDEX is
α if (+), γ if (−).

II. Y is N-S. INDEX is β.

III. Y is N-S. INDEX is β.

IV. O.B. is N-S. INDEX is
α if (+), γ if (−).

I. Assuming a N–S polarizer, the refractive index α would now be measurable from the crystal if it is (+) in sign whereas γ would be measurable if it is (−). A 45-degree counterclockwise rotation of crystal A would make the Y axis N–S; the resultant interference figure (II) indicates this by showing the optic plane to be vertical and to trend E–W. Thus the refractive index β could be measured from grain A in this orientation. Similarly, crystal B could be rotated 45 degrees counterclockwise to make its Y axis N–S and the refractive index β measurable from it; the resultant interference figure (III) shows the optic plane to be E–W and vertical for this orientation. Crystal C if rotated 45 degrees clockwise yields interference figure IV; this indicates that the plane perpendicular to its obtuse bisectrix is vertical and E–W. Its obtuse bisectrix is thus horizontal and N–S. Consequently, α or γ can be measured from the crystal, depending upon whether it is (+) or (−). Below interference figures I to IV in Fig. 10–4 is written (1) the identity of the principal vibration axis that is N–S and parallel to the plane of the stage and (2) the corresponding refractive index measurable from the crystal in this orientation (assuming a N–S polarizer).

MEASUREMENT OF 2V

The value of $2V$ can be calculated from the measured values of α, β, and γ or, more quickly, by use of the Mertie Chart (Fig. 9–6A) as already discussed (p. 159). However, for crystals of low birefringence, small experimental errors in the determination of α, β, and γ may produce a large error in the value of $2V$ thus determined. In an approximate manner $2V$ can be estimated from the curvature of the isogyres when the optic plane is at 45 degrees to the vibration directions of the polarizers (Fig. 10–5A). Usually, however, the following, more accurate methods of measuring $2V$ are utilized.

Fig. 10-4 (opposite). *Left-hand column:* Appearance of the interference figure for a crystal at 45 degrees off extinction as sketched on its upper surface. The optic plane (stippled) and the principal plane normal to the obtuse bisectrix (heavily shaded) are shown. For crystal *A*, both planes are perpendicular to the stage upon which it would rest, the obtuse bisectrix (O.B.) and the optic normal (Y) being parallel to the stage. (In the drawings the stage underlying the crystals is omitted.) For crystal *B*, only the optic normal is parallel to the stage, thus only the optic plane is perpendicular to the stage. In crystal *C*, only the obtuse bisectrix is parallel to the stage, thus only the heavily shaded plane is perpendicular to the stage. *Right-hand column, I-IV:* Appearance of the grain's interference figures if the crystals are rotated (see coiled arrows) so that the traces of the principal planes become parallel to the cross hairs; under each is given the refractive index measurable from the grain in that orientation (N-S polarizer assumed).

Mallard's Method

If two melatopes appear within the field of view, Mallard's method of determining $2V$ may be used. The pertinent equation is

$$D = K \sin E = K \beta \sin V$$

where D represents one-half the distance between the melatopes as measured in scale units of a micrometer eyepiece (Fig. 10–5B) and K represents a constant (commonly called Mallard's constant) whose value depends upon the numerical aperture of the objective and the number of micrometer scale divisions corresponding to a diameter of the visible field. For a given microscope, objective, and eyepiece, the value of K can be determined from crystals of known $2E$ (for example, barite, $2E = 63°12'$, $\beta = 1.636$; or aragonite, $2E = 30°52'$, $\beta = 1.681$) that are cut normal to the acute bisectrix. Assume Fig. 10–5B to represent the acute bisectrix figure of the

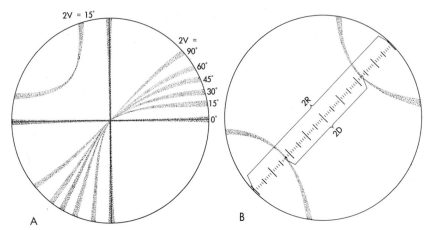

Fig. 10-5. (A) Determination of $2V$ from the curvature of the isogyres when the optic plane is at 45 degrees (NW-SE) to the polarizers. The curvature of the isogyres is illustrated for 15 degree intervals of $2V$ (After Wright, 1907, p. 338.) Both arms of the hyperbola are shown for $2V = 0°$ and $2V = 15°$. (B) Use of the micrometer ocular to measure $2D$ in Mallard's method and $2D/2R$ in Tobi's method.

barite plate; as illustrated, D measures as about 22.5 units. Mallard's constant for this microscope would then be

$$K = \frac{D}{\sin E} = \frac{22.5}{\sin 31° 36'} = 42.9$$

Henceforward, if the same microscope, objective, and micrometer eyepiece are used to measure D from the centered or near-centered acute bisectrix figure of an unknown, $2E$ or $2V$ can be calculated from the equations applicable to this microscope; that is,

$$\sin E = \frac{D}{K} = \frac{D}{42.9}$$

$$\sin V = \frac{D}{42.9 \; \beta}$$

For very precise work, Mallard's constant should be determined (for the microscope-objective-eyepiece combination to be used) by measurements on two or three different crystals whose known values of $2E$ differ widely.

Tobi's Method

The method of Tobi (1956, p. 516) eliminates the necessity of determining Mallard's constant, since twice the radius of the field of view (that is, $2R$) is measured in addition to $2D$, the intermelatope distance in Fig. 10–5B. If the numerical aperture of the objective used was 0.85, Fig. 10–6 permits $2E$ and $2V$ for the mineral to be determined from the ratio $2D/2R$. As inset A indicates, the value of $2D/2R$ locates a point on the vertical axis that also falls on a diagonal; this diagonal, when followed upward, indicates the value of $2E$ for the crystal measured. Similarly, inset B illustrates how the values of $2D/2R$ and of β represent coordinates of a point in the interior of the Tobi chart; this point falls on a diagonal that, if traced upward, indicates the value of $2V$ for this crystal.

Fig. 10–6 was constructed for an objective of N. A. 0.85. If an objective of a different numerical aperture (A') is used, the Tobi chart permits determination of $2E$ and $2V$ as before, provided the value $2D/2R$ measured with this objective is first multiplied by the ratio $A'/0.85$. If this objective will be customarily used, the foregoing step may be eliminated by multiplying all $2D/2R$ values in Fig. 10–6 by $0.85/A'$. With its vertical scale thus relabeled, Fig. 10–6 will be applicable to $2D/2R$ measurements made with this new objective.

Tobi states that the foregoing method may also be applied to off-centered acute bisectrix figures provided the trace of the optic plane passes through or near the center of the field of view. (Fig. 10–4B, if both melatopes were visible, would illustrate such a figure.) The intermelatope distance, $2D$, could be measured as for centered figures and the ratio $2D/2R$ would yield values of $2V$ with an error of 1 degree or less (Tobi, p. 519).

Other Methods

For centered interference figures in which both melatopes lie outside the field of view, $2V$ may be measured by the Kamb method (see p. 180). Commonly, however, the universal stage, an instrument that can be mounted on the microscope stage and permits a grain to be tilted into almost any desired position, is used to measure $2V$ with an accuracy of 1 to 2 degrees,

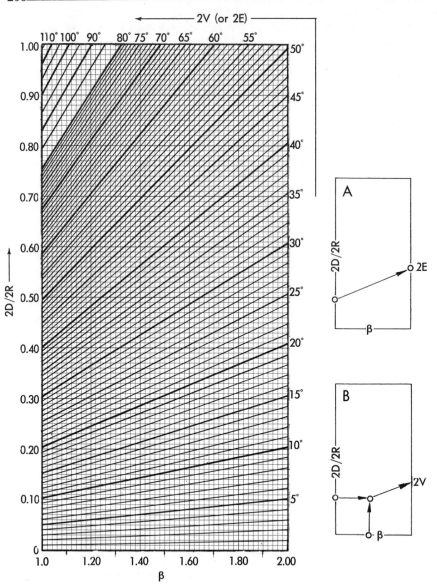

Fig. 10-6. Chart for determination of $2V$ or $2E$ in interference figures. (After Tobi, 1956.) Inset A shows use of the chart to determine $2E$ if $2D/2R$ has been measured; inset B indicates its use to determine $2V$ from the values of β and of $2D/2R$. In each case the point located by the measured values (encircled in insets A and B) falls on a diagonal line that indicates the value of the optic angle. Numerical aperture of objective is 0.85.

even if the figure is not centered. The procedures involved are beyond the scope of this textbook. Emmons (1943) discusses the method in great detail, and the attainable accuracy is discussed by Fairbairn and Podolsky (1951, p. 823) and by Wyllie (1959).

MEASUREMENT AND SIGNIFICANCE OF EXTINCTION ANGLES

The angular relations between a crystal's cleavage planes (or the crystal faces typically developed by its habit) and its X, Y, and Z directions determine both the extinction angle and the sign of elongation. In biaxial minerals three types of extinction exist: (1) *parallel* or *straight*, (2) *symmetrical*, and (3) *inclined* or *oblique*. The first two types have already been described for uniaxial crystals (p. 145) and need no further elaboration. The third type, inclined or oblique extinction, occurs if the grain's privileged directions are not parallel to the observed crystallographic features—that is, to its faces (Fig. 10–7A); to its cleavage planes (Fig. 10–7B); or to the traces of

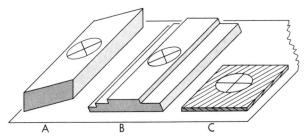

A   B   C

Fig. 10-7. Inclined extinction with respect to (A) the planes of an elongated crystal, (B) the cleavage planes in an elongated cleavage fragment, and (C) the traces of the cleavage planes in a thin section of the mineral. In (B) note that the superior (horizontal) cleavage is marked by the development of a larger surface area parallel to it.

its cleavage planes (Fig. 10–7C), if it is in a thin section. Consequently, if such crystals are at extinction, the observed crystallographic features will be at an angle T (the extinction angle) to the cross hairs. For pyroxenes and amphiboles, T is usually measured as the angle between the N privileged direction and the cleavage trace; for feldspars, it is measured as the angle between the n privileged direction and the cleavage trace. As a result, T may vary from 0 up to 90 degrees between different members of the pyroxene or amphibole groups, for example. Techniques of measuring T for biaxial minerals need no discussion since they are the same as for uniaxials (p. 145).

In crystallographic terminology a *zone* consists of a set of faces all of which, plus their lines of mutual intersection, are parallel to a given line (that is, the *zone axis*). Faces $(h\bar{k}0)$, (100), $(hk0)$, (110), and (010) in Fig. 10–8 belong to the same zone, the c axis being the zone axis. If a zone axis coincides with a principal vibration axis, X, Y, or Z, all faces in the zone will have a privileged direction (for normal incidence) parallel to this principal vibration axis (for example, the faces parallel to the c axis in Fig. 10–8A). Consequently, all faces in the zone will exhibit parallel extinction with respect to their lines of mutual intersection, since the latter are parallel to the zone axis. If the zone axis and principal vibration axis do not coincide, faces in the zone generally do not exhibit parallel extinction (with respect to their intersections). For example, of the planes parallel to the c axis in Fig. 10–8B, only (100), since it is also parallel to Y, shows

Fig. 10-8. Privileged directions for normal incidence upon the various faces in a zone parallel to the c axis if (A) c, the zone axis, coincides with a principal vibration axis (Z) or (B) the zone axis does not coincide with a principal vibration axis. (A) is the typical case for orthorhombic crystals; (B), the typical case for monoclinic crystals. The ellipses of intersection between the various faces and the crystal's indicatrix (the latter not drawn) are as shown, their major and minor axes being the privileged directions.

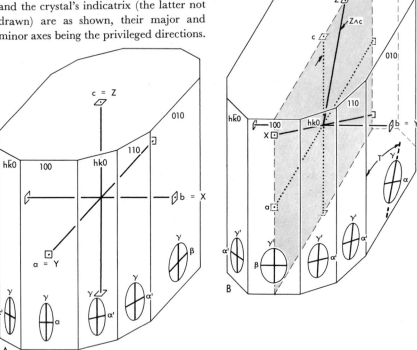

parallel extinction. The other faces exhibit inclined extinction, the extinction angle T increasing to a maximum as these planes approach parallelism to the (010) face. Since the (010) face is parallel to the shaded plane containing axes Z and c, the angle T for this face equals $Z \wedge c$ (read this symbol as "the angle between Z and c"). The angle T for the (010) face thus has a greater significance than the extinction angles for the other faces in Fig. 10–8B since the angle between a principal vibration axis and a crystallographic axis may be of use in identifying a mineral. Determination of $Z \wedge c$, for example, is helpful in the optical identification of individual varieties of pyroxene and amphibole.

In biaxial crystals the cleavage or crystal planes commonly observed generally occur in zones parallel to the crystallographic axes a, b, and c. Thus, in the orthorhombic system, since all three of these axes coincide with the principal vibration axes, parallel extinction occurs more frequently than inclined extinction (the latter being observable in orthorhombic crystals only for faces not parallel to a crystallographic axis). In monoclinic minerals, however, only the b axis coincides with one of the principal vibration axes. Thus parallel extinction can occur only for cleavages or crystal faces lying in a zone parallel to the b axis whereas inclined extinction occurs for faces in all other zones. Inclined extinction is therefore more frequent for monoclinic crystals than for orthorhombic. For triclinic crystals, since none of the crystallographic and principal vibration axes coincide, only inclined extinction is observed. By the same token, symmetrical extinction may be observed in orthorhombic crystals and, to a lesser extent, in monoclinic crystals but not in triclinic crystals.

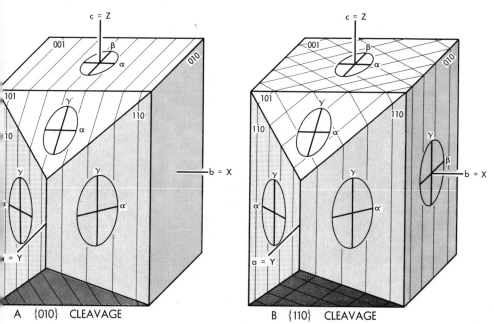

A {010} CLEAVAGE   B {110} CLEAVAGE

Fig. 10-9. Types of extinction for normal incidence of light upon various faces of an orthorhombic crystal if (A) the crystal possesses a cleavage direction that is parallel to (010) or (B) the crystal possesses two cleavage directions, (110) and (1Ī0). The traces of the cleavage on the other crystal faces and, in the case of (B), on each other, are drawn as fine lines.

As an example, consider an orthorhombic crystal with a {010} cleavage (Fig. 10–9A). Note that, with respect to the cleavage traces on each face, parallel extinction would be observed if the grain were lying on a plane parallel to (001), (101), (1Ī0), (110), (10Ī), etc. For an orthorhombic mineral possessing a prismatic {110} cleavage (Fig. 10–9B), the trace of cleavage planes (110) and (1Ī0), both on the other faces and on each other, indicates that parallel extinction would be observed on faces (1$\bar{1}$0), (110), and (010) whereas symmetrical extinction would be observed if the crystal were lying on (001), (101), or (10$\bar{1}$).

For a {110} cleavage in a monoclinic crystal (Fig. 10–10), parallel extinction would be observed only if the grain lay on (100); symmetrical, if on (001). If the grain lies on any other faces, inclined extinction occurs. Thus parallel and symmetrical extinction in monoclinic crystals is the exception rather than the rule. If the crystal in Fig. 10–10 possessed {010} cleavage instead of {110}, the trace of this {010} cleavage on the other planes could be drawn. The (100) and (001) planes would then exhibit parallel extinction, the others inclined. A face resting on (010) would then show no cleavage traces at all.

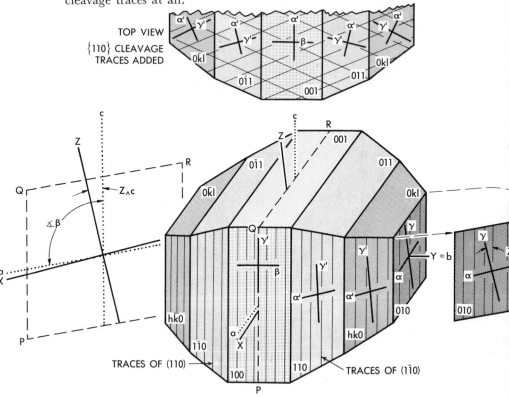

Fig. 10-10. Variation in the type of extinction for a monoclinic crystal possessing prismatic {110} cleavage according to the plane parallel to which it is cut (and rests upon). For a plate cut parallel to the (100) plane the crystal would exhibit parallel extinction; for one cut parallel to the (001) plane the crystal would exhibit symmetrical extinction. In the large majority of cases, however—for example, planes (110), (hk0), (010), (0kl), and (011)—inclined extinction would be observed, the extinction angle increasing for planes increasingly parallel to (010). For (010) this extinction equals the maximum value and corresponds to $Z \wedge c$ for the case illustrated. For simplicity the traces of the {110} cleavage on faces (0$\bar{k}$l), (0$\bar{1}$1), (001), (011), and (0kl) are omitted from the lower drawing.

In monoclinic crystals, the *b* axis coincides with a principal vibration axis (X, Y, or Z) for all wavelengths of light. The angle between axis *a* or *c* and the nearest principal vibration axis, however, varies slightly according to the wavelength of light used; $c \wedge Z$, for example, might differ slightly for different wavelengths. Thus, assuming that the trace of a cleavage parallel to the *c* axis occurs on a (010) plane, the extinction angle measured with respect to this trace might vary slightly according to the wavelength of light used as the illuminant.

Sign of Elongation

The identity of the principal vibration axis nearest the zone axis to which two or more cleavage or crystal planes are parallel can often be determined by observing the type of elongation for the crystals or fragments. Suppose the crystallographic features are two directions of almost equal ease of cleavage such that needles are developed in crushing the mineral. If the zone axis corresponds to X, the needles, regardless of which plane they are lying on, exhibit negative elongation (Fig. 10–11A); if it corresponds to Z, the needles exhibit positive elongation in all orientations (Fig. 10–11B); on the other hand, if the zone axis corresponds to Y, elongation may be positive or negative depending upon which plane the needle rests (Fig. 10–11C). Thus in a powder mount, if all needles exhibit negative elongation, X must be most nearly parallel to the needles' long axes; if all exhibit positive elongation, Z must be most nearly parallel to them; whereas if some needles exhibit positive and others negative elongation—that is, the elongation is ($\pm$)—Y must be most nearly parallel to their long axes.

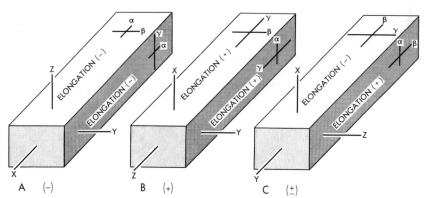

Fig. 10-11. Types of elongation in crystals (or cleavage fragments) of biaxial minerals. (A) Crystal is elongated parallel to X; therefore it is length fast (elongation negative) regardless of which face it rests upon. (B) Crystal is elongated parallel to Z; therefore it is length slow (elongation positive) regardless of which elongated face it rests upon. (C) Crystal is elongated parallel to Y and thus appears length fast (elongation negative) if resting on the plane as shown but appears length slow (elongation positive) if resting on the shaded face. Its elongation is thus denoted as ($\pm$).

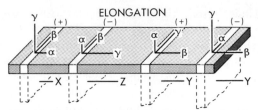

Fig. 10-12. Intersection between a rock thin section and plates (unshaded) of varying optic orientations.

For plate-shaped crystals or cleavage fragments, the principal vibration axis most nearly perpendicular to the plate can be determined from edge-on views. In powder mounts, since the grains rest on their flat surfaces, edge-on views are unattainable, but in thin sections cut through randomly oriented plates—for example, flakes of clay—such views are available. If X is perpendicular to the plates (as it is in the micas, glauconite, chlorites, vermiculite, and so forth), cross sections of the plates will always exhibit positive elongation. On the other hand, if Z is perpendicular to the plates, cross sections will always exhibit negative elongation. However, if Y is perpendicular to the crystal plates of that mineral, some sections will exhibit elongation $(+)$ and others elongation $(-)$, depending on whether the section through the plate is cut more nearly parallel to Z than to X (Fig. 10–12).

ABSORPTION AND PLEOCHROISM

With the identification of a particular axis, X, Y, or Z, as one of the privileged directions, the crystal is then oriented with this direction parallel to the polarizer in order to measure, as previously described, the principal index (α, β, or γ) associated with this direction. At the same time, however, the crystal's transmitted color for light vibrating parallel to X, Y, or Z (as the case may be) should also be observed and recorded. One variety of hypersthene, if in crystals about 0.03 mm thick, exhibits pleochroism as follows: X = red, Y = yellow, Z = blue. Obviously, the degree of absorption of the various wavelengths of white light in hypersthene varies according to whether the light is vibrating parallel to the X, Y, or Z principal vibration axes. To express the variations of absorption coefficients k_x, k_y, and k_z with wavelength of light in a manner similar to Fig. 8–23 would thus require three individual curves. To express the fact that, on the whole, the absorption of all light wavelengths is greatest for light vibrating, for example, parallel to Y, less if vibrating parallel to X, and least if vibrating parallel to Z, it is customary to write $A_y > A_x > A_z$ or simply $Y > X > Z$.

Vibration directions corresponding to the α' or γ' indices generally are associated with absorptions or transmitted colors intermediate between those for α and β or β and γ, respectively. For the hypersthene crystals cited above, for example, the color transmitted by an α' vibration direction might vary between orangish red and orangish yellow, according to how close the particular value of α' is to α or β, respectively. The transmitted color associated with γ', on the other hand, would vary from greenish blue (if γ' practically equals γ) to greenish yellow (if γ' only slightly exceeds β).

RECORDING DATA

A form for recording data—which also serves as a checklist for the optical properties that can be measured—appears in Appendix III (p. 285).

MINERAL IDENTIFICATION

If α, β, and γ have been accurately measured for an unknown biaxial mineral, the mineral species may usually be identified by consulting the figures and tables in Appendix II (pp. 255 ff.).

ELEVEN

Solution of Optical Problems by Use of Projections

As with all crystallographic problems involving only the angular relationships between lines and planes, many optical problems can be quickly solved by either the cyclographic projection or its close relative, the stereographic projection. Prior to their discussion, however, it is necessary to discuss the projection of lines and planes on a spherical surface, since both the cyclographic and stereographic projections are genetically related to projection on a sphere.

PROJECTION ON A SPHERE

Consider the following geometrical elements to pass through the center of a reference sphere (Fig. 11–1A): (1) a plane of random orientation, (2) a line (*OP*) perpendicular to this plane, and (3) a line (*OR*) within this plane. Their intersections with this sphere are, for the plane, a great circle* and, for each of the two lines, a point or, as it is called, a pole. Such intersections represent the *direct projections* of the given geometrical elements onto the sphere, the term "direct projection" indicating that the geometrical element itself was extended until it intersected the sphere.

In contrast, a given geometrical element may also be projected reciprocally. By definition, this signifies the erection of another geometrical element that is uniquely perpendicular to the given element; the intersection of this perpendicular element with the sphere represents the reciprocal

---

* A great circle may be defined as the arc of intersection of a sphere with a plane passing through its center.

(or polar) projection of the given element. For example, the geometrical element uniquely perpendicular to plane *A* in Fig. 11–1A is line *OP*. Consequently, line *OP*'s point of intersection with the sphere—that is, *face pole P*—represents the reciprocal or polar projection of plane *A*. Conversely, the geometrical element uniquely perpendicular to line *OP* is plane *A*. Consequently, the intersection of this plane with the sphere—that is, the dashed great circle in Fig. 11–1A—represents the reciprocal projection of line *OP*.

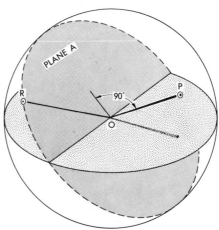

Fig. 11-1A. The dashed great circle and points *R* and *P* respectively represent the direct projection of plane *A*, line *OR*, and line *OP* on a reference sphere. Since plane *A* and line *OP* are mutually perpendicular, the dashed great circle could also represent the reciprocal (polar) projection of line *OP*; point *P* could be the reciprocal projection of plane *A*.

To summarize: (1) A plane or crystal face is directly projected on the sphere as a great circle and reciprocally projected as a face pole whereas (2) a line or crystal axis is directly projected as a pole and reciprocally projected as a great circle. (In either type of projection, the geometric element intersecting the sphere must necessarily pass through the sphere's center.)

THE CYCLOGRAPHIC
AND STEREOGRAPHIC PROJECTIONS

Relation to Spherical Projection

Unfortunately, a sphere cannot be printed on a flat page—a grave disadvantage for communication of ideas. However, several methods exist for transferring the points or great circles from a spherical surface to a flat plane. In the case of the cyclographic and stereographic projections, which are exceedingly convenient for solving problems in optical crystallography, this flat plane is an equatorial plane of the sphere (stippled in Fig. 11–1A). It is often called the plane of projection and divides the sphere into an upper and lower hemisphere. The line perpendicular to this equatorial plane (*UL* in Fig. 11–1B) intersects the sphere in an upper pole (*U*) and a lower pole (*L*).

A point on a hemisphere surface—for example, P in Fig. 11–1B—may be transferred to the equatorial plane by (1) constructing a line that joins this given point to the pole of the opposite hemisphere—for example, line LP—and (2) locating the intersection of this line—for example, P'—with the equatorial plane. Point P' marks the newly projected position of point P. A great circle on the sphere may be projected onto the equatorial plane by similarly treating individual points along it—for example, R and S—and obtaining their projected positions (R' and S').

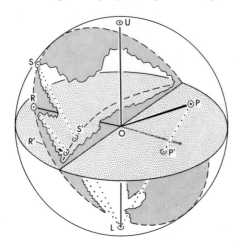

Fig. 11-1B. Transferral of points from the surface of the sphere to its equatorial plane (stippled). Each point is joined to the pole of the opposite hemisphere by a dotted line. This dotted line pierces the equatorial plane at the projected position of the point in this plane; R', S', and P' are examples. Point P' is the reciprocal projection of (shaded) plane A.

Point P' in Fig. 11–1B thus marks the projection in the equatorial plane of point P, which, in turn, may represent (1) the direct projection of line OP or (2) the reciprocal or polar projection of plane A. Depending upon its significance, therefore, point P' may represent the *cyclographic* projection of line OP (case 1) or the *stereographic* projection of plane A (case 2). The sole difference between these two projections is that the cyclographic stems from a direct projection whereas the stereographic stems from a reciprocal projection of a plane or line onto the sphere. Similarly, the dashed curve in the stippled equatorial plane of Fig. 11–1B may represent the cyclographic projection of plane A or the stereographic projection of line OP.

Coordinates of a Point

Face normal OP (which projects on the sphere at P and on the equatorial plane at P') becomes oriented in space if two angular coordinates, ϕ (phi) and ρ (rho), are defined for it (Fig. 11–2A). The value of ϕ indicates the angle between two planes; one (that is lightly shaded in Fig. 11–2A) contains line OP and axis UL, whereas the other (heavily shaded) is a previously selected reference plane that also hinges on UL. The value of ρ indicates the angle between OP and the polar axis UL. The value of ρ is defined to be positive if measured from the upper pole—for example, as for *OP*— and negative if measured from the lower pole—for example, as for *OQ*. By way of example (Fig. 11–2A), the coordinates of line *OP* are $\phi_1 = 45°$, $\rho_1 = 62°$ (approx.). Those for line *OQ* are $\phi = 135°$, $\rho = -65°$ (approx.).

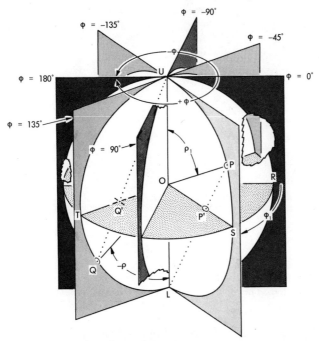

Fig. 11-2A. Spherical coordinate system for a sphere as related to pole *UL*. The shaded planes hinge on *UL* at various angles to the heavily shaded reference plane containing *R*. These angles represent the ϕ (phi) coordinates for these planes; negative ϕ values are measured counterclockwise from the reference plane, positive ϕ values clockwise. A line—for example, *OP*—is fixed in space if (1) the ϕ value of the plane containing it is specified and (2) its angle to pole *UL*—that is, its ρ (rho) angle—is also specified. In other words, if ϕ and ρ represent specific values, line *OP* is fixed in space.

Plotting Points

Let Fig. 11–2B represent Fig. 11–2A as seen by looking down the polar axis *UL*. Any radius of the circle in Fig. 11–2B may thus be considered to represent the edge-on view of a plane hinged on axis *UL*. The value of ϕ for such a plane can be measured with a protractor; it is the angle between this radius and *OR*, where *OR* in Fig. 11–2B represents the trace of the reference plane for which ϕ was assumed to equal 0 degrees. Positive ϕ angles are measured clockwise from *OR*; negative ones, counterclockwise. Radius *OS* in Fig. 11–2B, for example, represents the edge-on view of a plane that contains all directions (of which *OP* in Fig. 11–2A is one) for which ϕ equals 45 degrees. Similarly, radius *OT* represents the edge-on view of a plane that contains all directions for which ϕ equals 135 degrees.

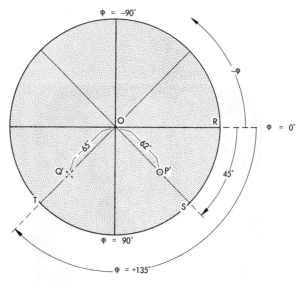

Fig. 11-2B. Stippled plane in (A) as seen looking down pole *UL*. The point *P'* has the coordinates $\phi = 45°$ and $\rho = 62°$.

Along such radii (for example *OS* in Fig. 11–2B) the distance *r*, at which a point (*P'*) plots outward from the circle's center, increases with ρ since

$$r = R \tan \frac{\rho}{2} \qquad \text{(Eq. 11–1)}$$

where *R* equals the radius of the reference sphere (*OS* in Fig. 11–2). Note that if ρ equals 0 degrees, *P'* plots at the circle's center (whereupon any value of ϕ pertains since *P'* then is common to all radii of the circle). If ρ equals 90 degrees, *P'* plots on the circumference. For values of ρ intermediate between 0 and 90 degrees, *P'* would plot somewhere along *OS*. The distance outward from the circle's center for a particular ρ value can be determined from any of the four radii that terminate at points *E*, *S*, *W*, and *N* of the stereographic (Wulff) net in Fig. 11–3, each radius being divided into 2-degree intervals of ρ.

THE STEREONET

Description and Significance

The series of curves joining points *N* and *S* of the stereonet (Fig. 11–3) represent great circles—that is, the cyclographic projection of a family of

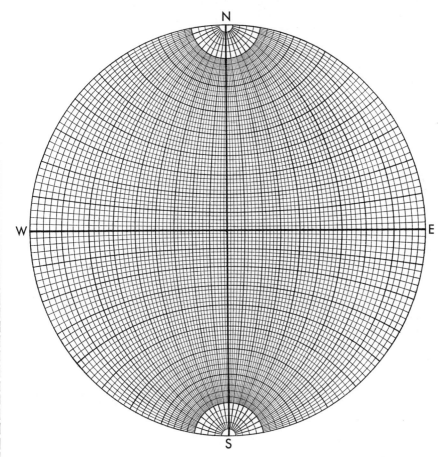

Fig. 11-3. The stereonet (Wulff net) drawn with 2-degree intervals.

planes that hinge on line *NS* (Fig. 11–4A). The angular interval between them is actually 2 degrees, but they are shown only at 15-degree intervals in Fig. 11–4A. With one exception, these planes have been omitted from

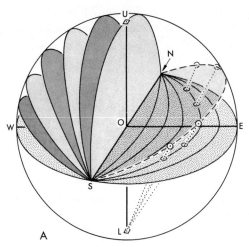

Fig. 11-4. (A) Significance of the great circles joining points *N* and *S* in Fig. 11-3. (B) Significance of the small circles in Fig. 11-3. If the shaded plane is rotated on *NS* as a hinge, the intersections of the lines within it (*OA, OB, ..., OH*) with the reference sphere describe small circles (dashed) on the sphere. If all points on these small circles are transferred to the equatorial plane by the standard stereographic method (as shown for two points on small circle *B*) the small circles of the stereonet result.

the upper right quadrant of the sphere to show better their cyclographic projections on the equatorial plane. The dashed great circle in which the transparent plane intersects the sphere is shown projected upon the equatorial plane in detail.

A second set of curves on the stereonet is roughly transverse to the great circles just described. To understand their significance, assume semicircle SFN (Fig. 11–4B) to contain 90 equally spaced radii, a few of which—for example, OS, OA, OB, OC, etc.—are drawn in Fig. 11–4B. If semicircle SFN is rotated around hinge line NS, its radii trace out a series of small circles* (dashed) on the sphere surface. If each point on these small circles is projected onto the stippled plane, the solid curves shown in Fig. 11–4B result.

Practical Use

Remove Fig. 11–3 from the book and mount it on a square of heavy cardboard; transfix its precise center with a thumbtack whose point has been ground very sharp and thin (Fig. 11–5A). On a sheet of tracing paper draw two, mutually perpendicular, intersecting lines. Pierce their intersection with a needle, label their ends $\phi = 0°$, $90°$, $-90°$, and $180°$, and impale this tracing paper on the stereonet's tack (Fig. 11–5B), using the previously made needle hole as a guide. To plot a point P ($\phi_1 = 30°$, $\rho_1 = 75°$) on the tracing paper, rotate the paper above the stereonet until the stereonet's radius OE makes an angle of $30°$ ($= \phi_1$) with the $\phi = 0°$ line on the paper. Looking through the tracing paper, count off $75°$ ($= \rho_1$) outward along radius OE and mark the point thus located (that is, point P) with a small pinhole in the tracing paper.

If this point P represents the stereographic projection of a crystal plane whose spherical plot is on the upper hemisphere, draw a tiny circle around the pinhole (both to indicate what P is and to make the pinhole easier to find). If point P represents a face projected onto the lower hemisphere (in which case its coordinates would be $\phi_1 = 30°$ and $\rho_1 = -75°$), draw a small cross over the pinhole instead of encircling it.

To plot the cyclographic projection of a crystal plane, if its ϕ and ρ coordinates are known, simply plot its face pole P in the foregoing manner. Point P will now overlie diameter WE of the stereonet (Fig. 11–5B). Now count off 90 degrees from P along stereonet diameter WE, as seen through the tracing paper. The great circle that intersects WE at the point thus located may now be traced onto the tracing paper (see heavy curve NMS in Fig. 11–5B) to mark the cyclographic projection of the plane whose face pole was P.

A pole indicative of a direction or axis (rather than of a face normal)

---

* A small circle may be defined as any circle on a sphere's surface whose radius is less than that of the sphere.

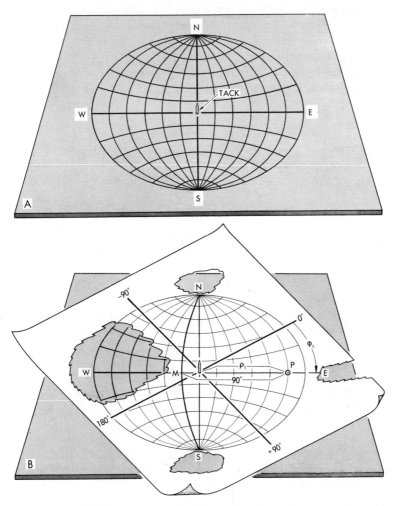

Fig. 11-5. (A) The stereonet (shaded gray for illustrative purposes) made ready for use by insertion of a fine-pointed tack through its center. (B) Insertion of the stereonet's tack through the intersection of two perpendicular lines on a sheet of tracing paper. The ends of the lines are labeled with their ϕ values (0°, 90°, 180°, and — 90°). The paper has been rotated above the stereonet until diameter WE of the underlying stereonet makes an angle of ϕ with the paper's 0-degree line, the circumference of the stereonet serving as a protractor. A point P, whose coordinates are ϕ_1 and ρ_1, may then be plotted on the overlying paper by counting a number of degrees equal to ρ_1 outward from the tack toward E. If P represents the face pole of a stereographically plotted crystal plane, point M, which is 90 degrees from P on diameter WE, marks the great circle NMS that is the cyclographic projection—that is, trace—of this crystal plane.

within a crystal would be cyclographically plotted from its ϕ and ρ coordinates just as face pole P was plotted (p. 220). In this case, however, a tiny square should be drawn around the pinhole to distinguish this cyclographic projection of a direction or axis from the stereographic projection of a crystal face.

Given the cyclographic projection on a piece of tracing paper of two directions within a crystal (for example, Q and R in Fig. 11–5C), the

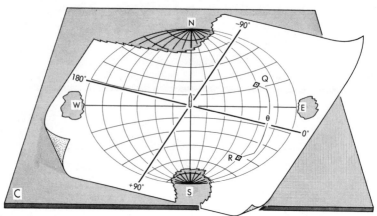

Fig. 11-5C. The tracing paper has been rotated until two points, Q and R, lie on the same great circle of the underlying stereonet. This permits θ, the angle between Q and R, to be measured by counting the degree divisions (on the great circle of the underlying stereonet) between Q and R.

cyclographic projection of the crystal plane that either contains or is parallel to both these directions is obtained by rotating the tracing paper so that both points overlie the same great circle. This great circle (curve $NQRS$ in Fig. 11–5C), if traced onto the tracing paper, then represents the desired crystal plane. The reason for this is obvious since the cyclographic projection of a line—for example, OR in Fig. 11–1—necessarily falls upon the cyclographic projection of the plane that contains it—for example, plane A in Fig. 11–1.

Similarly, θ, the angle between two directions cyclographically projected as points Q and R (in Fig. 11–5C), can be measured by rotating the underlying stereonet until both points fall along the same great circle. The portion of this great circle lying between points Q and R is divided (by its intersections with the small circles) into 2-degree intervals on the stereonet. Counting the number of these intervals between points Q and R determines the value of θ. If Q and R had represented face poles, θ would denote the interfacial angle between the crystal faces they represented.

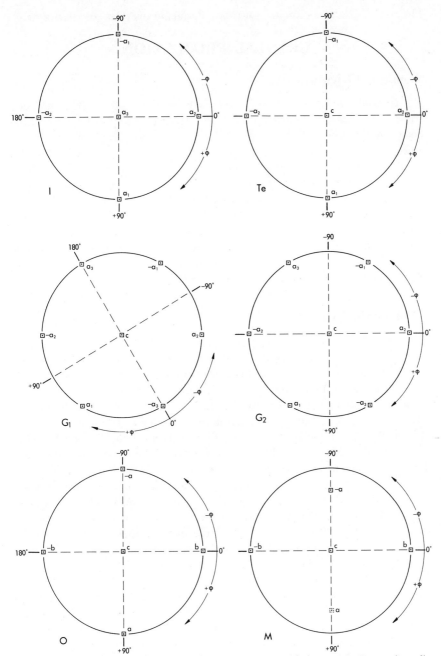

Fig. 11-6. Standard coordinates for the crystal axes (cyclographically projected) in the isometric system (I), tetragonal system (Te), hexagonal system (G₁ or G₂ alternatives), orthorhombic system (O), and monoclinic system (M). The standard coordinates for the positive end of the a axis in monoclinic crystals are $\phi = 90°$ and $\rho = \beta - 180°$. Since β is an obtuse angle, this positive end therefore projects upon the lower hemisphere as shown in Fig. 11-7A. To indicate its position on the lower hemisphere, the cyclographic projection of the positive a axis is marked with a dashed square rather than a solid square.

CRYSTAL PROJECTION

Standard Orientations

In projecting the faces or axes of a crystal, the crystallographic axes are assigned a standard orientation with respect to the sphere of projection. For example, a crystal's vertical axis (a_3 in isometric, c in all other systems) must always coincide with the sphere's polar axis (UL in Fig. 11–2A). Similarly, its a or a_1 axis (except for hexagonal crystals) is always made to lie within the $\phi = +90°$ plane. Fig. 11–6 illustrates the standard orientations of the crystal axes in stereograms for all but the triclinic system. There are two standard orientations (usually called G_1 and G_2) for the hexagonal system. Many leading American crystallographers prefer the G_1 position whereas the noted German crystallographer V. Goldschmidt used the G_2 position, sometimes exclusively, in his publications. The G_2 position seems more analogous to the standard positions for the other systems.

Monoclinic Example*

Fig. 11–7A illustrates the projection on the fundamental sphere of the axes and faces of a monoclinic crystal; the faces are projected reciprocally by the dashed lines (normal to them) that emanate from the sphere's center and terminate at their intersection with the sphere's surface. In Fig. 11–7B these intersections have been transferred to the equatorial plane by joining them to the pole of the opposite hemisphere—U or L, as the case may be. The method of transferral (previously discussed on p. 216) is shown in detail for face poles (011) and (01$\bar{1}$) and both ends of the a axis, the imaginary lines joining them to the opposite pole being drawn as dotted lines. Faces or lines that were originally projected upon the upper hemisphere are respectively symbolized with tiny circles or squares; faces or lines originally on the lower hemisphere are respectively symbolized as tiny crosses or dashed squares. Similarly, a great circle on the upper hemisphere—for example, the upper half of the dot-dashed great circle in Fig. 11–7A—will be shown by a dot-dashed line after transferral to the equatorial plane; one on the lower hemisphere—for example, the lower half of the dot-dashed great circle in Fig. 11–7A—will be shown, after transferral to the equatorial plane, by a line composed of three dots alternating with dashes. Fig. 11–7C is the equatorial plane of Fig. 11–7B as it would look when tilted into the plane of the paper; it is therefore the stereogram of the crystal shown in Fig. 11–7A. Note that the b axis and the normals to faces (1$\bar{1}$0), (110), (010), and ($\bar{1}$10) all lie within the equatorial plane (in Fig. 11–7A) and thus intersect the sphere at points along its equator. Consequently, these points retain their same positions after transferral to the equatorial plane.

---

* The reader is assumed to understand Miller indices, (hkl), etc., and axial ratios, $a: b: c$. If he is not familiar with these, he should consult an elementary textbook on mineralogy or crystallography.

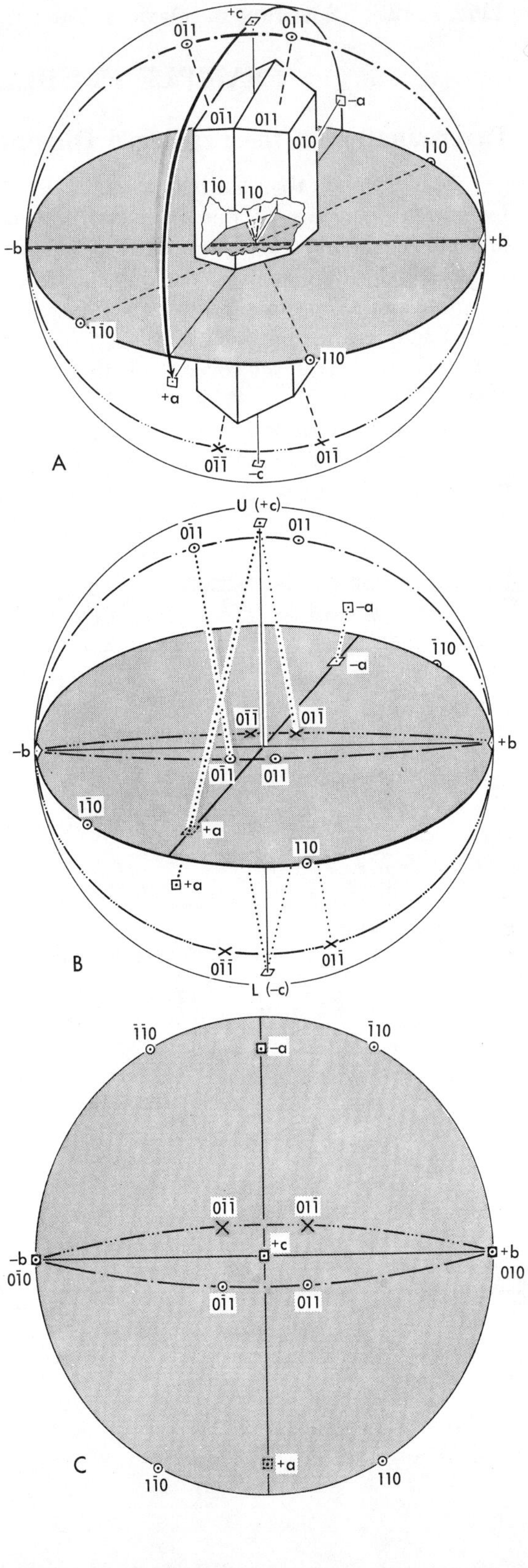

Fig. 11-7. (A) Spherical projection of a monoclinic crystal. The dashed lines normal to the various faces emanate from the sphere's center and intersect the sphere's surface where shown. (B) Transferral of the spherically projected points to the equatorial plane is shown in detail for planes (0$\bar{1}$1) and (01$\bar{1}$) and both ends of the *a* axis. (C) View of the equatorial plane as seen looking down *UL*—that is, the "stereogram" of the crystal.

SAMPLE PROBLEMS

Determination of the Privileged Directions for a Crystal Plane

Given plane (hkl) of a biaxial mineral. The two privileged directions for normal incidence on this plane can be determined if the following are known: (1) the angular relationships between its crystallographic axes and its principal vibration axes, (2) its acute bisectrix and value for $2V$, and (3) the ϕ and ρ coordinates of the given face. Palache, Berman, and Frondel (1951) summarize items (1), (2), and (3) for a large number of minerals. For minerals not as yet covered by this reference, items (1) and (2) are given by Winchell and Winchell (1951). In cases where item (3) is not available in print, the ϕ and ρ coordinates for a given face may be calculated from its Miller indices (hkl) and from the axial ratio $a : b : c$ by means of the formulas for the tangents of ϕ and ρ (summarized in Table 11–1).

Table 11–1

FORMULAS FOR COMPUTATION OF TAN ϕ AND ρ FOR A FACE FROM ITS MILLER INDICES (hkl) AND FROM THE AXIAL RATIO, $a : b : c$

Tan ϕ	Tan ρ	
	$k \neq 0$	$k = 0$
Monoclinic		
$\dfrac{hc/a + l \sin (\beta - 90)}{kc \sin \beta}$	$\dfrac{kc}{l \cos \phi}$	$\dfrac{hc/a + l \sin (\beta - 90)}{l \sin \phi \cos (\beta - 90)}$
Orthorhombic		
$\dfrac{h}{ak}$	$\dfrac{kc}{l \cos \phi}$	$\dfrac{hc}{la \sin \phi}$
Tetragonal		
$\dfrac{h}{k}$	$\dfrac{kc}{l \cos \phi}$	$\dfrac{hc}{l \sin \phi}$
Hexagonal (G_2)		
$1.1547 \left(\dfrac{h}{k} + 0.5 \right)$	$\dfrac{kc}{l \cos \phi}$	$\dfrac{hc}{l \sin (\phi - 30)}$

We want, for example, to determine the privileged directions for a plate of augelite, $Al_2(PO_4)(OH)_3$, cut parallel to its (111) plane. The following data are obtained from Palache, Berman, and Frondel (1951, p. 871):

(1) $X = b$, $Y_{\wedge} c = -56°$, $Z_{\wedge} c = +34°$, $\beta = 112° \ 27'$

(2) biaxial $(+)$, $2V = 51°$

(3) stereographic coordinates for (111), $\phi = 52° \ 37\frac{1}{2}'$ and $\rho = 46° \ 18\frac{1}{2}'$

In (1) the negative sign conventionally indicates that a principal vibration axis—Y, in this case—occurs within the *acute* angle between the a and c axes; the positive sign indicates Z to occur within the *obtuse* angle β between the a and c axes. Winchell and Winchell (1951) use these signs in just the opposite manner.

Utilizing the foregoing data, the crystal axes (a, b, and c), the principal vibration axes (X, Y, and Z), the optic axes (A and B), and the (111) plane have been plotted cyclographically in Fig. 11-8. All except the last are lines and thus plot cyclographically as points. The XZ (optic plane) is cyclographically projected as a dashed great circle along which, $25\frac{1}{2}$ degrees in either direction from Z, the optic axes are located. Since the optic axes are perpendicular to the circular sections of the indicatrix, the cyclographic projection of both circular sections may be located by using points A and B in the manner previously discussed (p. 220). The resultant great circles

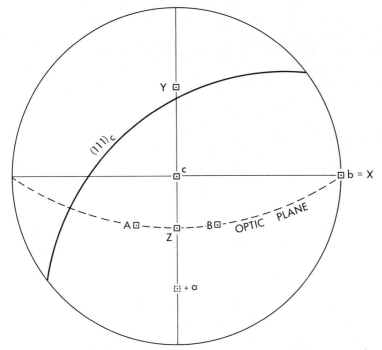

Fig. 11-8. Cyclographic projection of the X, Y, Z, and a, b, c axes and the (111) face for augelite.

(dotted and labeled C.S.A. and C.S.B. in Fig. 11–9) intersect the cyclographic projection of (111) at points L_A and L_B. These points represent the cyclographic projections of two lines in plane (111) that are also radii of the circular sections of the indicatrix. Thus, if plane (111) intersected the indicatrix for augelite, the intersection would be an ellipse for which lines L_A and L_B would represent two radii of equal length, both equaling β (see inset in Fig. 11–9). One of the two semiaxes of this ellipse bisects the angle between lines L_A and L_B (since the bisector of the angle between two equally long radii of an ellipse necessarily coincides with a semiaxis). The cyclographic projection of one of the semiaxes of this ellipse thus lies at Z', midway (in terms of degrees) between L_A and L_B on great circle $(111)_c$. If a great circle of the underlying stereonet is made to coincide with $(111)_c$, one can then count 90 degrees from point Z' to locate X', the cyclographic projection of the other semiaxis of the ellipse. Semiaxes Z' and X' of the ellipse of intersection between plane (111) and the indicatrix are the privileged directions for normal incidence of light on (111); in Fig. 11–9 the one closer to Z is labeled Z', the one closer to X is labeled X'.

Determination of Extinction Angles

Augelite has a perfect {110} cleavage, both directions of which, (110) and ($\bar{1}$10), would intersect the (111) plane. Thus, a plate of augelite grown or cut parallel to the (111) plane would show these intersections as fine cracks or lines. The extinction angles between these cracks and the Z' privileged direction for augelite lying on a (111) plane are determined by cyclographically plotting cleavage planes (110) and ($\bar{1}$10), using the ϕ coordinates given by Palache et al. (1951, p. 871). The resultant great circles—actually dot-dashed straight lines in Fig. 11–9—intersect great circle $(111)_c$ at points m_1 and m_2. These points represent the cyclographic projection of the lines of intersection of the {110} cleavages with plane (111)—that is, the cyclographic projection of the cleavage traces on (111). The angles between Z' and m_1 and between X' and m_2, if counted in degrees along great circle $(111)_c$, are approximately 23 and 7 degrees, respectively. These angles thus represent the extinction angles for augelite lying on a (111) face.

The procedure for determining the extinction angles of a crystal with respect to a particular cleavage as it lies on a face (hkl) is as follows: (1) Determine the privileged directions for face (hkl), (2) locate the cleavage traces on (hkl), and (3) measure the angle between these cleavage traces and the privileged directions.

Further information about the optical properties of a (111) plate of augelite may be obtained from Fig. 11–9. For example, by counting the degrees between m_1 and m_2, the angle between the cleavage traces on a (111) plate is determined to be about 60 degrees. Furthermore, the interference figure from such a plate would be a nearly centered optic axis figure since optic axis B and the normal to the (111) plate almost coincide.

Specifically, the angle between optic axis B and the (111) face pole in Fig. 11–9 is 6 degrees. Light traveling along the normal to (111) emerges at the center of the field in an interference figure from a grain resting on a (111) plane. Thus light rays that traveled along optic axis B emerge fairly close to the center; applying the Tobi method to the melatope that marks their point of emergence, these rays would be found to have traveled (while in the crystal) at an angle of 6 degrees with respect to the normal to the microscope stage.

Calculation of Refractive Index

Since the privileged directions X' and Z' for normal incidence on an (hkl) plane have already been located, the grain's refractive indices for light vibrating parallel to these directions can be calculated by using Eq. 9–2 (p. 155), provided the ϕ and ρ coordinates of X' and of Z' are known for the crystal in "standard optical orientation"—that is, with Z at $\rho = 0°$

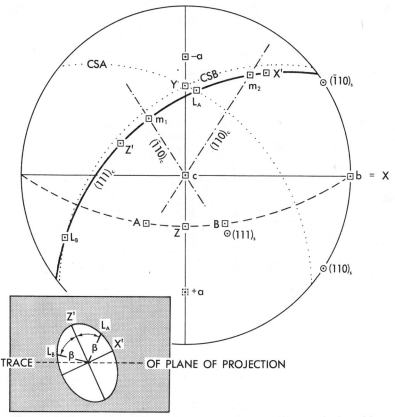

Fig. 11-9. Use of the cyclographic projection in solving optical problems. The optical and crystallographic directions for augelite are plotted. The subscripts s and c denote the stereographic and cyclographic projections, respectively. For further details of symbolism see text.

and with X at $\phi = 0°$ and $\rho = 90°$. Continuing with augelite as an example, we note that Fig. 11–9 is a standard crystallographic orientation (see Fig. 11–6M). Thus, as Fig. 11–10 shows, the c axis is perpendicular to the plane of projection. To cause the Z axis to become perpendicular thereto requires a 34-degree rotation of the entire crystal about its b axis. During this rotation the projected positions of Z, Y, X', and Z' would move along the hollow arrows to their new positions. Note that the direction labeled X' in Fig. 11–10 can be extended downward so as to intercept the lower hemisphere as well as the upper. After a 34-degree rotation about the b axis, as shown by the coiled arrow in Fig. 11–10, this line $X'X'$ becomes $X'_r X'_r$ and now intercepts the upper hemisphere on its forward surface.

In practice, these new positions, after this rotation, are obtained as follows (Fig. 11–11): (1) Rotate the original projection—Fig. 11–9—on the tack that pierces it until Z, the direction to be rotated to the center, falls along an E–W diameter of the stereonet. (2) The number of degrees between Z and the center is the amount the crystal must be rotated to cause Z to take position Z_r. (3) The same rotation of the crystal that caused Z to move to Z_r causes the other points (X', Z', Y,) to move 34 degrees along small circles of the underlying stereonet to their rotated positions (X'_r, Z'_r, etc.). Since we continue to plot all directions on the upper hemisphere only, the path that direction X' traces out during the 34-degree rotation is com-

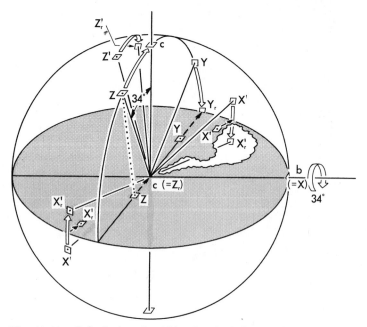

Fig. 11-10. Spherical and, within the shaded plane, cyclographic projection of the optic directions X, Y, Z, X', and Z' to show their loci as the entire crystal is rotated 34 degrees around the b axis. After completion of this rotation, their positions are X'_r, Z_r, Y_r, and so forth.

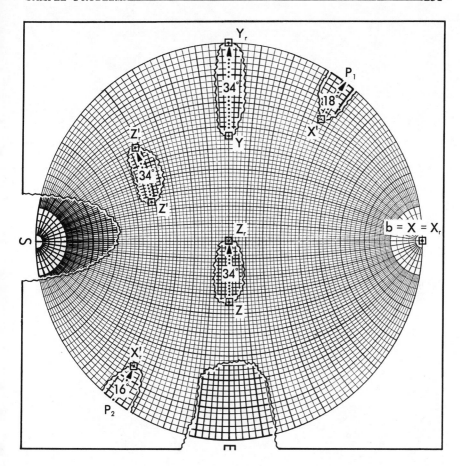

Fig. 11-11. Use of the stereonet to obtain the rotated positions of points. The stereonet is rotated below the tracing paper until Z, the direction desired to be rotated to the center, lies above diameter WE of the stereonet. The 34-degree rotation of the crystal illustrated in Fig. 11-10 then causes Z to move to Z_r, Z' to move 34 degrees along a small circle to Z'_r, and X' to move 34 degrees along a small circle to X'_r.

plicated by the fact that after 18 degrees of this rotation (which makes $X'X'$ horizontal) we plot the path of its other end for the last 16 degrees of rotation. Note that when X' was horizontal, its one end plotted at P_1 and its other at P_2.

The ϕ and ρ coordinates of X'_r and of Z'_r, if substituted into Eq. 9–2 along with the values of the principal refractive indices of augelite ($\alpha = 1.5736$; $\beta = 1.5759$; $\gamma = 1.5877$), permit α' and γ' for a grain lying on the (111) plane to be calculated as 1.5756 and 1.5765, respectively.

REFERENCES CITED IN TEXT

BAUER, N. (1949). In *Physical methods of organic chemistry* (A. Weissberger, ed.). Vol. 1, Pt. 2, pp. 1203-1214.

BIOT, J. B. (1820). Mémoire sur les lois générales..., *Mém. Acad. France, Année 1818*, III, pp. 177-384.

BOUMA, B. J. (1947). *Physical aspects of colour*, Eindhoven (Netherlands): N. V. Philips, 312 pp.

DITCHBURN, R. W. (1952). *Light*, London: Blackie and Son, 680 pp.

EMMONS, R. C. (1943). The universal stage, *Geol. Soc. Amer. Mem.*, 8, 205 pp.

EMMONS, R. C. and GATES, R. M. (1948). The use of Becke line colors in refractive index determination, *Am. Mineral.*, *33*, pp. 612-618.

FAIRBAIRN, H. W. and PODOLSKY, T. (1951). Notes on precision and accuracy of optic angle determination with the universal stage, *Am. Mineral.*, *36*, pp. 823-832.

FISHER, D. JEROME (1958). Refractometer perils, *Am. Mineral.*, *43*, pp. 777-780.

FLETCHER, L. (1891). The optical indicatrix and the transmission of light in crystals, *Min. Mag.*, *9*, p. 278.

FOSTER, W. R. (1947). Gravity-separation in powder mounts as an aid to the petrographer, *Am. Mineral.*, *32*, pp. 462-467.

FRESNEL, A. (1827). Mémoire sur la double réfraction, *Mém. Acad. France, Année*, VII, pp. 45-176.

HARDY, A. C. and PERRIN, F. H. (1932). *The principles of optics*, New York: McGraw-Hill, 632 pp.

JOHANNSEN, ALBERT (1918). *Manual of petrographic methods*, New York: McGraw-Hill, 649 pp.

KAMB, W. BARCLAY (1958). Isogyres in interference figures, *Am. Mineral.*, *43*, pp. 1029-1067.

LARSEN, ESPER S. and BERMAN, HARRY (1934). The microscopic determination of the nonopaque minerals. *U. S. Geol. Surv.*, Bull. 848.

LONGHURST, R. S. (1957). *Geometrical and physical optics*, London: Longmans, Green, 534 pp.

LOUPEKINE, I. S. (1947). Graphical derivation of refractive index ϵ for the trigonal carbonates, *Am. Mineral.*, *32*, pp. 502-507.

MERTIE, John B., Jr. (1942). Nomograms of optic angle formulae, *Am. Mineral.*, *27*, pp. 538-551.

PALACHE, CHARLES, BERMAN, HARRY, and FRONDEL, CLIFFORD (1951). *Dana's system of mineralogy* (7th ed.), Vol. II, New York: John Wiley and Sons, 1124 pp.

RINNE, F. and BEREK, MAX (1953). *Anleitung zu optischen Untersuchungen mit dem Polarisationsmikroskop*, Stuttgart: E. Schweizerbart, 366 pp.

ROSENFELD, JOHN L. (1950). Determination of... difficultly oriented minerals..., *Am. Mineral.*, *35,* pp. 902-905.

SAYLOR, CHARLES P. (1935). Accuracy of microscopical methods for determining refractive index by immersion, *J. Res. Nat. Bur. Stand.*, *15*, pp. 277-294.

SHAUB, B. M. (1959). Using the microscope for specific gravity determination of minute mineral grains, *Am. Mineral.*, *44*, pp. 890-891.

TOBI, A. C. (1956). A chart for measurement of optic axial angles, *Am. Mineral.*, *41*, pp. 516-519.

WEISS, C. S. (1817). Über ein verbesserte method für die Bezeichnung der verschiedenen Flächen eines Krystallisationssystems, *Abh. Akad. Wiss. Berlin, Physik. Kl.*, 1816-1817, pp. 231-285.

WILCOX, RAY E. (1959). Use of the spindle stage for determination of principal indices of refraction..., *Am. Mineral.*, *44*, pp. 1272-1293.

WINCHELL, A. N. (1931). *Microscopic characters of artificial minerals*, New York: John Wiley and Sons, 403 pp.

WINCHELL, A. N. and WINCHELL, HORACE (1951). *Elements of optical mineralogy*, New York: John Wiley and Sons, 551 pp.

WRIGHT, F. E. (1913). The index ellipsoid (optical indicatrix) in petrographic microscope work, *Am. Jour. Sci.*, *185*, pp. 133-138.

WRIGHT, F. E. (1923). Interference figures..., *Jour. Opt. Soc. Amer.*, 7, pp. 779-817.

WRIGHT, W. D. (1958). *The measurement of color*, New York: The Macmillan Company, 263 pp.

WYLLIE, P. J. (1959). Discrepancies between optic axial angles of olivines measured over different bisectrices, *Am. Mineral.*, *44*, pp. 49-64.

APPENDIX I ___ PROPERTIES OF ELLIPSES

As shown in Fig. I–1 an ellipse possesses a major axis, AA', and perpendicular to it a minor axis, BB'; these are respectively its longest and shortest diameters. One half these distances, labeled a and b, are called the

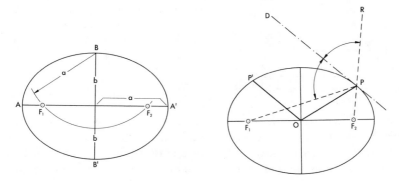

Fig. I-1 (left). Location of the two foci, F_1 and F_2, in an ellipse whose semi-axes are a and b. An arc of length a is swung with B as its center. Its points of intersection with major axis AA' locate F_1 and F_2.

Fig. I-2 (right). Construction of OP', the radius of the ellipse conjugate to OP. Dashed construction lines were drawn from foci F_1 and F_2 through P. Line DP, which bisects the angle between these dashed construction lines, corresponds in direction to OP'.

semimajor and semiminor axes. The two foci, F_1 and F_2, are always on the major axis and may be located by swinging an arc of radius a (centered on B, the end of the minor axis) so that it intersects the major axis. The focal distances, OF_1 or OF_2 in Fig. I–2, may also be calculated from the pythagorean formula

$$OF_1 = OF_2 = \sqrt{a^2 - b^2}$$

Radii extending outward from O, the ellipse center, are called simply radii; those extending outward from one of the focal points are called focal radii. A radius OP' (Fig. I–2) is said to be conjugate to a given radius OP if OP' is parallel to the tangent to the ellipse at P. The direction of this tangent (DP in Fig. I–2) may be determined by (1) extending one of the focal radii (F_2P) through point P to a distant point such as R and then (2) bisecting the angle F_1PR. Then OP', the radius conjugate to OP, is the direction through O parallel to DP, the bisector of angle F_1PR.

The construction of an ellipse is sometimes necessary in the graphical solution of optical problems. A convenient method for constructing an ellipse of major and minor axes $2a$ and $2b$, respectively, is the rectangular method wherein two concentric circles of radius a and b are drawn on finely lined graph paper (Fig. I–3). The major and minor axes are then

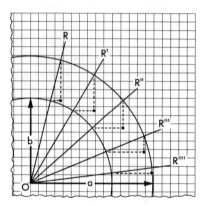

Fig. I-3. Rectangular method of constructing an ellipse beginning with circles whose radii correspond to the ellipse's semiaxes, a and b. More closely spaced points along the ellipse can be obtained by drawing more numerous construction radii (OR, OR', ...). Only a quadrant of the ellipse is shown.

drawn outward from O, the common center of the two circles, each axis being parallel to the rulings of the graph paper. A series of random lines— OR, OR', and so on—are also drawn outward from O. At each line's intersection with the major circle, a dashed construction line parallel to the minor axis is drawn; at its intersection with the minor circle, a dashed construction line parallel to the major axis is drawn. These two dashed lines intersect at a point on the ellipse. Generally, in optical problems, only a small portion of the ellipse need be constructed.

APPENDIX II___ $\quad$ DETERMINATIVE OPTICAL
TABLES AND CHARTS

The tables and charts in this section permit, in many cases, the identification of a mineral if its optical properties are known. The data used are chiefly those of Larsen and Berman (1934), to which further additions from the literature have been made. For the anisotropic minerals a series of charts has been devised on the basis of which the identity of an unknown mineral can be narrowed down in minutes to, in most cases, a relatively few possibilities. These charts can be used not only in the powder-immersion method of optical mineralogy but also in thin-section determinations since birefringence [that is, $(\epsilon - \omega)$ for uniaxials and $(\gamma - \alpha)$, $(\beta - \alpha)$, or $(\gamma - \beta)$ for biaxials] is plotted. After a tentative identification of a mineral has been made, it is advisable to check further all facts known about the specimen with such detailed descriptions of the optical properties of the suspected identity as are compiled in Larsen and Berman (1934) or Winchell and Winchell (1951).

ISOTROPIC MINERALS

The list supplied in Table II–1 permits determination of an unknown isotropic mineral from its refractive index (n) and, if necessary, its specific gravity (G). The latter value may be determined, when needed, by the sink-float technique (p. 63) using the refractive index media as heavy liquids (Fig. 5–14).

Table II–1

ISOTROPIC MINERALS

(in order of increasing refractive index up to $n = 1.999$)

n	G	Color*	Mineral name
1.326	2.79	R	Villiaumite
1.328	2.79	R	Villiaumite
1.339	2.75	Gy	Hieratite
1.339	2.78	Gy	Cryolithionite
1.370	2.0	—	Cryptohalite
1.403	1.21?	W, Va	Termierite
1.406	1.9–2.3	Va	Opal
1.427	2.61	C, W, Y	Ralstonite
1.434	1.86	C	Opal
1.434	3.36–3.63	Vi, Bl	Yttrocerite
1.434	3.18	C, Pu, Va	Fluorite
1.44±	2.1	Va	Opal
1.44±	2.5	Br, Bk	Hisingerite
1.454	2.49	C	Sulphohalite
1.456	1.76	C	Alum
1.457	3.55	Y	Yttrofluorite
1.459	1.64	W	Tschermigite
1.46	2.2	Va	Opal
1.461	2.04	C	Melanophlogite
1.462	2.19	C	Lechatelierite
1.47±	2.6	Br, Bk	Neotocite
1.47±	1.8	Bl–Gn	Allophane
1.479	2.26~	W, Gy, Gn	Analcite
1.48	1.92	W	Faujasite
1.48	2.3	W, C, Gy	Noselite
1.483	2.30	Gy, Bl	Sodalite
1.485	1.94	C	Evansite
1.486	2.3	W	Cristobalite
1.487	2.25	W, Gy, Gn	Analcite
1.487	3.32	R–Vi	Hackmanite
1.49	1.86	Bl, Gn	Allophane
1.490	1.99	C	Sylvite
1.495	2.3	Bl	Noselite
1.496	2.4	Bl	Haüynite
1.496	—	Gn, W	Bolivarite

* The abbreviations for the colors are the same as those given by Larsen and Berman (1934) and are as follows: Amb, amber; Bk, black; Bl, blue; Br, brown; C, colorless; Dk, dark; Gn, green; Gy, gray; O, orange; Pi, pink; Pu, purple; R, red; Va, variable; Vi, violet; W, white; and Y, yellow. Hyphenated symbols such as Gn-Bl should be read greenish blue. Although not explicitly stated by Larsen and Berman (1934), the colors in this column are evidently reflected colors, not transmitted.

Table II–1 (continued)
Isotropic Minerals

n	G	Color	Mineral name
1.50±	2.4	Bl	Lazurite
1.50±	—	—	Stevensite
1.5±	2.2	Y	Rosiéresite
1.505	1.98	W, Y, Gn	Vashegyite
1.508	2.46–2.59	C	Tychite
1.509	2.47	C	Leucite
1.51	—	W, Va	Cimolite
1.51	2.5–3	Br, Bk	Hisingerite
1.51	2.5	Bl, Gn	Haüynite
1.514	2.38	C	Northupite
1.517	2.65	Gn	Planerite
1.517	⟨1.0	W	Meerschaum
1.518	—	C	Pollucite
1.525	2.9	C	Pollucite
1.53±	2.34	W	Kehoeite
1.533	2.83	C	Langbeinite
1.535	1.07	Amb–Y	Succinite
1.54±	2.7	Br, Bk	Neotocite
1.54±	—	Bl–Gn	Cornuite
1.541	1.05	Y, R–Br	Ajkaite
1.542	1.09	—	Telegdite
1.542	2.6	W	Halloysite
1.544	2.17	C	Halite
1.555	2.6	W	Halloysite
1.558	—	—	Serpentine (Ni)
1.565	—	Bl	Traversoite
1.569	2.6	W	Collophanite
1.57±	2.6	Gn	Zaratite
1.57±	2.5–3	Br–Bk	Hisingerite
1.572	3.25	C	Nitrobarite
1.572	3.02	R	Manganolangbeinite
1.58±	2.3–2.5	Y	Koninckite
1.584	1.95–2.05	W	Schroetterite
1.59±	3	Bk, Br–Bk	Hisingerite
1.59±	2.6±	Gn	Zaratite
1.59±	2.7±	W	Collophanite
1.59	2.6±	Gn	Garnierite
1.590	2.93	—	Kochite
1.596	2.88	C	Zunyite
1.60±	5±	Gy, Y, C	Stibiconite

Table II–1 (continued)

ISOTROPIC MINERALS

n	G	Color	Mineral name
1.60±	2.7±	R–Br	Borickite
1.600	2.88	C	Zunyite
1.602	2.79	Gn, Br, Bk	Voltaite
1.605	2.97	Br	Grodnolite
1.61±	2.6±	Gn	Zaratite
1.61±	2.03	Br, Y	Diadochite
1.61	2.96	Br, Bk	Allanite
1.63±	2.9±	W	Collophanite
1.635	2.5	Br, Y, W	Pitticite
1.638	—	Pi–Br	Hydrothorite
1.639	1.53	C	Sal ammoniac
1.64	2.83	Dk–Br	Picite
1.64±	3.4	Br	Griphite
1.640	3.35	Bk	Homilite, metamict
1.65±	2–3	Y	Koninckite
1.65	2.66	Y–Br	Egueiite
1.65±	4.2	Dk–Br	Tritomite
1.65	2.8	Gn, Y, Br	Greenalite
1.653	3.32	Br	Lovchorrite
1.66	3.40	Br	Griphite
1.66±	3	Br, Bk	Hisingerite
1.67±	2.7±	R–Br	Borickite
1.67	3.05	C	Hibschite
1.674	3±	Gn, Y, Br	Greenalite
1.675	3.0	Br, Gn	Plazolite
1.68±	3.4	Br, Bk	Allanite
1.69	3.40	C	Rhodizite
1.69	5.2–5.4	Gn, O, Bk, Br	Orangite
1.70±	5.0	Gy	Stibiconite
1.70	5.00	Dk–Br, Bk	Polycrase
1.702	4.12	Gn	Arandisite
1.705	3.510	R	Pyrope
1.707	3.95	Y, O–Y	Berzeliite
1.710	4.414	—	Uranothorite
1.715	3.58	Br, R, Bl, Gn, Y, Bk	Spinel
1.716	1.99(?)	Br	Delvauxite
1.718	3.6±	R	Spinel
1.720	3.683	Dk–Gn	Spinel
1.72±	3.5–4.2	Br, Bk	Allanite
1.725	4.52	Gn–R	Rowlandite

Table II–1 (continued)
Isotropic Minerals

n	G	Color	Mineral name
1.735	3.530	Va	Grossularite
1.736	3.6	C	Periclase
1.739	3.2	Y	Helvite
1.74±	4.6±	—	Pilbarite
1.74±	4.30	Br	Caryocerite
1.742	3.715	R	Pyrope
1.748	4.27	Y, O–Y	Berzeliite
1.75±	3.75	Bk, Gn	Spinel
1.754	3.43	R	Danalite
1.755	3.70	W	Arsenolite
1.757	4.2	Dk–Br	Tritomite
1.758	4.58	Bk, Br, Gn	Yttrialite
1.760	3.837	Dk–R	Rhodolite
1.763	3.633	Br	Hessonite
1.766	—	R	Almandite
1.77±	4.13	Br, R	Melanocerite
1.77	5.44	Bk	Mackintoshite
1.77±	3.8±	Bk	Pleonaste
1.779	4.45	Y, O–Y	Berzeliite
1.780	4.0–4.5	Bk, Gn–Bk	Gadolinite
1.782	4.38	Gy–Gn	Gahnite
1.789	3.917	Dk–R	Almandite
1.790	—	Bl	Gahnite
1.80±	3.9	Bk	Hercynite
1.800	4.180	C, R	Spessartite
1.80±	5±	Gy, Y	Stibiconite
1.801	4.903	Dk–R	Almandite
1.802	—	Gy–Bk	Gahnite
1.805	4.117	R	Spessartite
1.805	4.55–4.62	C, Gn	Gahnite
1.812	4.14	Br, Y	Beckelite
1.815	4.211	R	Spessartite
1.818	—	R	Almandite
1.818	4.09	Dk–Gn, Br	Naegite
1.82	4.12	Gn	Arandisite
1.826	4.0–4.3	C	Malacon
1.83	5.04	Y	Romeite
1.83	4.39	Bk	Hercynite
1.830	4.250	Dk–R	Almandite
1.838	3.32	C	Lime

Table II–1 (*continued*)

ISOTROPIC MINERALS

n	G	Color	Mineral name
1.838	3.418	Gn	Uvarovite
1.84	3.75	Gn	Uvarovite
1.86±	4.8±	Gy, Gn	Bindheimite
1.86	3.78	Gn	Uvarovite
1.865	3.781	Va	Andradite
1.87±	5.07	Y	Romeite
1.87±	3.77	Gy–Br	Chalcolamprite
1.88	4.3–4.55	Bk	Tscheffkinite
1.88	3.76	Bk, R, Br	Andradite
1.89±	3.6–3.9	Dk–Br, Bk	Ellsworthite
1.895	3.750	Y, Gn, Br, Bk	Andradite
1.90±	5	Gy, Y	Stibiconite
1.92±	4	Gn–Bk	Betafite
1.923	4.23	Bk	Galaxite
1.925	5.51	Y, Br, R	Microlite
1.93	3.93	C	Nantokite
1.935	5.5–5.7	Gn	Microlite
1.94	3.7	Bk	Melanite
1.94	5.24	Y	Samirésite
1.96±	5.19	Y	Neotantalite
1.96	4.3±	Br	Pyrochlore
1.965	4.3–4.55	Bk	Tscheffkinite
1.97	5.8	Br	Djalmaite
1.98±	4.8±	Br	Hatchettolite
1.98	3.85	Bk	Schorlomite
1.99±	5	Y, Gy	Stibiconite
1.99±	5.95	Y–Br	Microlite

UNIAXIAL MINERALS

Optical data for 342 uniaxial mineral specimens are summarized in Table II–2. The first part of the table lists positive minerals; the second part, negative minerals. Each uniaxial mineral is assigned a code number (left-hand column). In Figs. II–1 and II–2, which follow the table, these minerals are plotted with respect to ω and the absolute value of their birefringence, $(\epsilon - \omega)$. Consequently, if ϵ and ω have been measured for an unknown mineral, the point with coordinates ω and $(\epsilon - \omega)$ may be located on Fig. II–1 or II–2, depending upon whether the crystal is positive or negative. A square with sides equal to 0.004—that is, twice the approximate determinative error of refractive index measurement—can be centered on this coordinate point. The code numbers within this square then indicate the minerals in Table II–2 that may be identical with the unknown. If several possible choices are indicated, a reference text (Winchell and Winchell, 1951) should be consulted for more detailed descriptions (cleavage and so forth) with which the unknown may be more closely compared and hence ultimately identified. If the unknown uniaxial mineral is not one of the uniaxial minerals listed in Table II–2, the charts cannot, of course, be used for its identification.

Figs. II–1 and II–2 may be of use in thin-section studies wherein the optic sign and the birefringence, $(\epsilon - \omega)$ or $(\omega - \epsilon)$, are determinable. Thus, if the birefringence and optic sign are known, they define a vertical, bar-shaped area in one of the figures; this area is likely to contain, among others, the code number that will identify the unknown mineral. Of course, since ω is unknown, there will in general be a number of choices. Visual estimates of ω through observation of the mineral's relief with respect to Canada balsam—or to a neighboring mineral of known refractive index—will reduce this multiplicity.

Table II–2

UNIAXIAL MINERALS

Code number	ω	ϵ	Mineral name
Optically Positive			
1.	1.46	1.54	Chrysocolla
2.	1.46±	1.57	Chrysocolla
3.	1.461	1.474	Tincalconite
4.	1.47	1.50	Hatchettite
5.	1.470	1.474	Gmelinite
6.	1.475	1.486	Laubanite
7.	1.480	1.482	Chabazite
8.	1.487	1.492	Aphthitalite
9.	1.488	1.500	Douglasite
10.	1.490	1.502	Natrolite hydronephelite
11.	1.490	1.496	Aphthitalite
12.	1.500	1.510	Liskeardite
13.	1.508	1.509	Leucite
14.	1.515	1.54	Ozocerite
15.	1.518	1.522	Leifite
16.	1.518	1.521	Davyne
17.	1.52	1.55	Koenenite
18.	1.521	1.529	Microsommite
19.	1.522	1.527	Natrodavyne
20.	1.533	1.575	Fibroferrite

Table II–2 (*continued*)
UNIAXIAL MINERALS

Code number	ω	ε	Mineral name
		Optically Positive	
21.	1.535	1.537	Apophyllite
22.	1.536	1.572	Coquimbite
23.	1.536	1.572	Ashcroftine
24.	1.544	1.553	Quartz
25.	1.550	1.650	Vaterite
26.	1.550	1.556	Coquimbite
27.	1.556	1.645	Julienite
28.	1.559	1.627	Ferrinatrite
29.	1.559	1.580	Brucite
30.	1.562	1.576	Chlorite, white
31.	1.565	1.575	Pinnoite
32.	1.566	1.585	Brucite
33.	1.572	1.592	Alunite
34.	1.576	1.579	Penninite
35.	1.576	1.584	Ekmannite
36.	1.580	1.588	Coeruleolactite
37.	1.582	1.645	Cacoxenite
38.	1.583	1.602	Alumian
39.	1.589	1.590	Rinneite
40.	1.590	1.600	Wardite
41.	1.59	1.60	Manganbrucite
42.	1.597	1.612	Amesite
43.	1.604	1.615	Sarcolite
44.	1.606	1.611	Eudialyte
45.	1.609	1.630	Narsarsukite
46.	1.614	1.648	Pseudowollastonite
47.	1.619	1.627	Pseudowavellite
48.	1.620	1.654	Churchite
49.	1.620	1.630	Hamlinite (Goyazite)
50.	1.622	1.631	Pseudowavellite
51.	1.623	1.625	Metatorbernite
52.	1.626	1.640	Svanbergite
53.	1.629	1.639	Hamlinite (Goyazite)
54.	1.630	1.640	Deltaite
55.	1.630	1.640	Pseudowavellite
56.	1.632	1.639	Akermanite
57.	1.633	1.639	Akermanite
58.	1.634	1.640	Hamlinite (Goyazite)
59.	1.644	1.697	Dioptase
60.	1.654	1.670	Phenacite

Table II–2 (continued)
Uniaxial Minerals

Code number	ω	ε	Mineral name
		Optically Positive	
61.	1.654	1.676	Plumbogummite
62.	1.654	1.703	Rhabdophanite
63.	1.655	1.708	Dioptase
64.	1.662	1.663	Cahnite
65.	1.665	1.685	Auerlite, altered
66.	1.671	1.689	Hinsdalite
67.	1.676	1.757	Parisite
68.	1.680	1.685	Florencite
69.	1.691	1.719	Willemite
70.	1.697	1.719	Willemite
71.	1.701	1.726	Willemite
72.	1.714	1.732	Troostite
73.	1.716	1.721	Idocrase
74.	1.717	1.818	Bastnaesite
75.	1.719	1.733	Bromellite
76.	1.720	1.827	Xenotime
77.	1.721	1.816	Xenotime
78.	1.724	1.746	Connellite
79.	1.730	1.810	Mixite
80.	1.75±	1.765	Buttgenbachite
81.	1.750	1.752	Abukumalite
82.	1.757	1.804	Benitoite
83.	1.778	1.801	Coniçhalcite
84.	1.794	1.803	Arseniopleite
85.	1.80	1.81	Thorite
86.	1.90	2.12	Trippkeite
87.	1.910	1.945	Ganomalite
88.	1.913	1.923	Nasonite
89.	1.918	1.934	Scheelite
90.	1.920	1.937	Scheelite
91.	1.926	1.985	Zircon
92.	1.936	1.991	Zircon
93.	1.945	1.971	Nasonite
94.	1.948	1.958	Hedyphane
95.	1.96	2.02	Zircon
96.	1.973	2.656	Calomel
97.	1.974	1.984	Powellite
98.	1.997	2.093	Cassiterite
99.	2.001	2.097	Cassiterite
100.	2.013	2.029	Zincite

Table II–2 (continued)
UNIAXIAL MINERALS

Code number	ω	ε	Mineral name

Optically Positive

Code number	ω	ε	Mineral name
101.	2.114	2.140	Phosgenite
102.	2.13	2.21	Penfieldite
103.	2.19	2.21	Kleinite
104.	2.21	2.22	Iodyrite

Optically Negative

Code number	ω	ε	Mineral name
105.	1.465	1.464	Gmelinite
106.	1.480	1.478	Chabazite
107.	1.481	1.461	Hanksite
108.	1.487	1.484	Cristobalite
109.	1.487	1.486	Analcite
110.	1.488	1.474	Ettringite
111.	1.490	1.471	Loewite
112.	1.496	1.491	Gmelinite
113.	1.496	1.491	Levynite
114.	1.500	1.464	Thaumasite
115.	1.507	1.468	Thaumasite
116.	1.507	1.500	Sulphatic cancrinite
117.	1.509	1.486	Nocerite
118.	1.512	1.487	Nocerite
119.	1.512	1.498	Hydrotalcite
120.	1.516	1.470	Beidellite
121.	1.520	1.512	Tachyhydrite
122.	1.524	1.496	Cancrinite
123.	1.530	1.506	Slavikite
124.	1.532	1.528	Nepheline
125.	1.532	1.527	Kaliophilite
126.	1.532	1.529	Milarite
127.	1.533	1.530	Chalcedony
128.	1.534	1.514	Zinc aluminate
129.	1.534	1.522	Dipyre (4.2% K_2O)
130.	1.537	1.535	Apophyllite
131.	1.537	1.533	Nepheline
132.	1.539	1.511	Mellite
133.	1.539	1.530	Chalcedony
134.	1.539	1.535	Nephelite

Table II–2 (continued)
UNIAXIAL MINERALS

Code number	ω	ε	Mineral name
Optically Negative			
135.	1.540	1.536	Marialite
136.	1.540	1.510	Brugnatellite
137.	1.542	1.538	Nephelite
138.	1.542	1.516	Stichtite
139.	1.545	1.503	Pholidolite
140.	1.547	1.522	Fluoborite
141.	1.549	1.536	Gyrolite
142.	1.550	1.542	Dipyre
143.	1.550	1.540	Dipyre
144.	1.552	1.543	Dipyre
145.	1.554	1.541	Dipyre
146.	1.558	1.545	Dipyre
147.	1.560	1.540	Jefferisite
148.	1.560	1.495	Trudellite
149.	1.565	1.560	Zeophyllite
150.	1.565	1.545	Dipyre
151.	1.566	1.528	Fluoborite
152.	1.567	1.545	Wernerite
153.	1.568	1.564	Beryl
154.	1.569	1.550	Dipyre
155.	1.570	1.545	Mizzonite
156.	1.575	1.552	Mizzonite
157.	1.575	1.550	Mizzonite
158.	1.575	1.57	Calcioferrite
159.	1.576	1.546	Parsettensite
160.	1.579	1.577	Penninite
161.	1.580	1.553	Mizzonite
162.	1.581	1.551	Mizzonite
163.	1.581	1.575	Beryl
164.	1.582	1.555	Mizzonite
165.	1.582	1.545	Mizzonite
166.	1.583	1.549	Mizzonite
167.	1.584	1.554	Mizzonite
168.	1.585	1.551	Mizzonite
169.	1.586	1.560	Uranospinite
170.	1.587	1.336	Soda niter
171.	1.587	1.559	Mizzonite
172.	1.588	1.553	Mizzonite
173.	1.590	1.560	Mizzonite
174.	1.59±	1.56	Connarite

Table II–2 (*continued*)

UNIAXIAL MINERALS

Code number	ω	ε	Mineral name
		Optically Negative	
175.	1.591	1.573	Metavoltine
176.	1.592	1.582	Torbernite
177.	1.594	1.556	Mizzonite
178.	1.595	1.557	Mizzonite
179.	1.598	1.590	Beryl
180.	1.600	1.563	Meionite
181.	1.60±	1.55±	Chloraluminite
182.	1.600	1.585	Dehrnite
183.	1.601	1.591	Dennisonite
184.	1.603	1.598	Dahllite
185.	1.607	1.571	Meionite
186.	1.612	1.593	Meliphanite
187.	1.613	1.607	Fluocerite
188.	1.618	1.552	Chalcophyllite
189.	1.620	1.609	Dahllite
190.	1.621	1.619	Gillespite
191.	1.621	1.618	Eucolite
192.	1.621	1.611	Lewistonite
193.	1.623	1.613	Dehrnite
194.	1.623	1.620	Merrillite
195.	1.624	1.613	Lewistonite
196.	1.626	1.605	Bazzite
197.	1.627	1.582	Troegerite
198.	1.629	1.624	Francolite
199.	1.630	1.622	Podolite
200.	1.632	1.626	Melilite
201.	1.632	1.602	Bementite
202.	1.632	1.575	Chalcophyllite, altered
203.	1.633	1.629	Melilite
204.	1.633	1.629	Voelckerite
205.	1.633	1.630	Fluorapatite, pure
206.	1.634	1.629	Fluorapatite
207.	1.635	1.631	Dahllite
208.	1.636	1.613	Dravite
209.	1.637	1.615	Mitscherlichite
210.	1.639	1.637	Strontianapatite
211.	1.640	1.636	Wilkeite
212.	1.640	1.633	Dehrnite
213.	1.641	1.621	Calcium tourmaline, uvite
214.	1.642	1.623	Rubellite

Table II–2 (continued)
UNIAXIAL MINERALS

Code number	ω	ε	Mineral name
			Optically Negative
215.	1.643	1.623	Zeunerite
216.	1.646	1.641	Manganianfluorapatite
217.	1.646	1.625	Indicolite
218.	1.647	1.636	Gehlenite
219.	1.648	1.625	Rubellite
220.	1.649	1.644	Apatite
221.	1.649	1.643	Daphnite
222.	1.650	1.646	Wilkeite
223.	1.650	1.624	Bementite
224.	1.650	1.575	Szaibelyite
225.	1.651	1.644	Hydroxylapatite
226.	1.653	1.623	Friedelite
227.	1.655	1.650	Wilkeite
228.	1.655	1.651	Manganapatite
229.	1.655	1.650	Ellestadite
230.	1.655	1.645	Koninckite
231.	1.655	1.633	Tourmaline
232.	1.657	1.653	Gehlenite
233.	1.658	1.486	Calcite, pure
234.	1.661	1.657	Manganianhydroxylapatite
235.	1.662	1.633	Tourmaline
236.	1.662	1.636	Tourmaline
237.	1.664	1.629	Friedelite
238.	1.666	1.658	Gehlenite
239.	1.667	1.664	Chlorapatite
240.	1.667	1.490	Plumbocalcite
241.	1.669	1.658	Gehlenite
242.	1.670	1.658	Gehlenite
243.	1.671	1.639	Schorlite
244.	1.672	1.662	Hardystonite
245.	1.672	1.644	Tourmaline
246.	1.675	1.636	Pyrosmalite
247.	1.675	1.59	Chloromagnesite
248.	1.679	1.502	Magnesiodolomite
249.	1.680	1.655	Sincosite
250.	1.681	1.647	Schorlite
251.	1.681	1.643	Schallerite
252.	1.681	1.500	Dolomite, pure
253.	1.682	1.662	Tourmaline
254.	1.684	1.672	Svabite

Table II–2 (*continued*)
UNIAXIAL MINERALS

Code number	ω	ε	Mineral name
		Optically Negative	
255.	1.686	1.505	Magnesiodolomite
256.	1.687	1.641	Chromium tourmaline
257.	1.690	1.673	Ferroakermanite
258.	1.69	1.60	Stilpnomelane
259.	1.691	1.691	Gehlenite
260.	1.694	1.641	Spangolite
261.	1.698	1.658	Schorlite
262.	1.698	1.518	Ankerite
263.	1.700	1.509	Magnesite
264.	1.704	1.697	Schallerite
265.	1.705	1.701	Idocrase (Vesuvianite)
266.	1.706	1.698	Svabite
267.	1.707	1.698	Genevite
268.	1.712	1.700	Idocrase, Be-bearing
269.	1.713	1.705	Idocrase, cyprine
270.	1.716	1.526	Ankerite
271.	1.718	1.700	Ferroschallerite
272.	1.719	1.715	Idocrase
273.	1.721	1.534	Calcite, Mn-bearing
274.	1.723	1.681	Pyrochrorite
275.	1.726	1.527	Magnesite
276.	1.728	1.531	Ankerite
277.	1.73±	1.72	Melanocerite
278.	1.733	1.714	Hematolite
279.	1.736	1.732	Idocrase
280.	1.743	1.546	Mangandolomite
281.	1.748	1.645	Freirinite
282.	1.749	1.547	Ankerite
283.	1.76	1.63	Stilpnomelane
284.	1.760	1.557	Cordylite
285.	1.765	1.555	Ferrodolomite
286.	1.768	1.760	Corundum
287.	1.772	1.763	Dark red corundum, ruby
288.	1.772	1.770	Swedenborgite
289.	1.777	1.772	Britholite
290.	1.788	1.570	Mesitite
291.	1.800	1.750	Ammoniojarosite
292.	1.80	1.72	Ferritungstite
293.	1.815	1.761	Molybdophyllite
294.	1.816	1.728	Borgstroemite

Table II–2 (continued)
Uniaxial Minerals

Code number	ω	ε	Mineral name
Optically Negative			
295.	1.816	1.597	Rhodochrosite
296.	1.816	1.600	Rhodochrosite
297.	1.817	1.597	Rhodochrosite
298.	1.820	1.715	Jarosite
299.	1.82	1.73	Carphosiderite
300.	1.826	1.605	Rhodochrosite
301.	1.830	1.745	Ammoniojarosite
302.	1.830	1.750	Natrojarosite
303.	1.830	1.596	Siderite
304.	1.849	1.621	Smithsonite
305.	1.849	1.615	Manganosiderite
306.	1.850	1.625	Smithsonite
307.	1.853	1.803	Hoegbomite
308.	1.855	1.613	Siderite
309.	1.855	1.600	Sphaerocobaltite
310.	1.870	1.792	Arseniosiderite
311.	1.87	1.85	Dussertite
312.	1.872	1.634	Siderite
313.	1.875	1.633	Siderite
314.	1.875	1.784	Plumbojarosite
315.	1.882	1.785	Argentojarosite
316.	1.898	1.815	Arseniosiderite
317.	1.905	1.785	Argentojarosite
318.	2.01	1.99	Armangite
319.	2.01	1.82	Bismite
320.	2.026	2.010	Hedyphane
321.	2.03	2.00	Pseudoboléite
322.	2.041	1.926	Cumengeite
323.	2.050	2.042	Pyromorphite
324.	2.05	2.03	Boléite
325.	2.07	2.05	Barysilite
326.	2.09	1.94	Hydrocerussite
327.	2.13	1.94	Bismutosphaerite
328.	2.135	2.118	Mimetite
329.	2.15	2.04	Matlockite
330.	2.16	2.14	Bellite
331.	2.25	2.20	Endlichite
332.	2.26±	2.10	Hetaerolite
333.	2.269	2.182	Stolzite
334.	2.295	2.285	Finnemanite

Table II–2 (*continued*)

UNIAXIAL MINERALS

Code number	ω	ε	Mineral name
		Optically Negative	
335.	2.31	1.95	Geikielite
336.	2.32Li	2.25Li	Ecdemite
337.	2.34±	2.14	Hetaerolite
338.	2.35Li	2.33Li	Lorettoite
339.	2.354	2.299	Vanadinite
340.	2.36Li	2.25Li	Schwartzembergite
341.	2.36Li	2.31Li	Langbanite
342.	2.40	2.28	Wulfenite

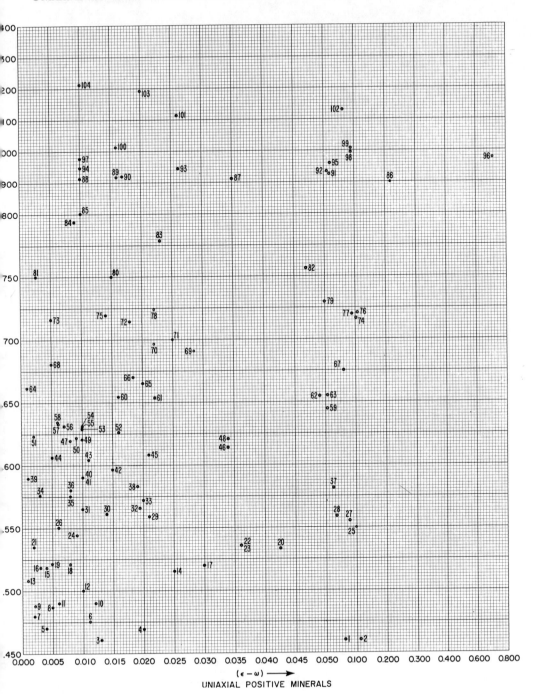

Fig. II-1. Determinative chart for uniaxial positive minerals.
Consult Table II-2 to identify code numbers.

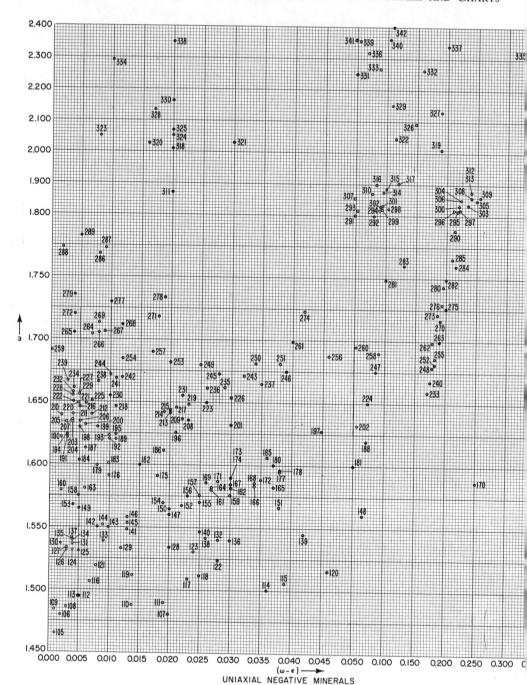

Fig. II-2. Determinative chart for uniaxial negative minerals.
Consult Table II-2 to identify code numbers.

BIAXIAL MINERALS

Tables II–3 and II–4 respectively list the positive and negative biaxial minerals in order of increasing value of β; the minerals are assigned code numbers (left-hand column) in numerical sequence. The biaxial positive minerals are plotted in Figs. II–3 to II–5, $(\gamma - \alpha)$ and $(\beta - \alpha)$ locating the coordinates of the point where the mineral's code number is written. The code numbers of the biaxial negative minerals are located in Figs. II–6 to II–8 at points with coordinates $(\gamma - \alpha)$ and $(\gamma - \beta)$. The sloping lines indicate different values of $2V$ and are accurate to about 3 to 5 degrees. Thus, if the values α, β, and γ, and the optic sign, have been measured for an unknown biaxial mineral, these data define a point on one of the graphs. A square whose sides equal 0.004, if centered on this point, will probably contain the code number for the mineral's identity, assuming it is not an extremely rare mineral. Several code numbers will appear within the square. They should be recorded on scrap paper, although many of them can be immediately eliminated if they are numbers that do not fall within the same "Range of β" as does the measured value of β. For convenience, these ranges are summarized within the insets at the upper left of Figs. II–3 to II–8. The code numbers will thus usually narrow down to a relatively few candidates. Comparison of the detailed properties of these candidate minerals (obtainable from Winchell and Winchell, 1951) with those of the unknown will often permit an unequivocal identification. Such additional comparative data might include specific gravity, extinction angles, sign of elongation, and dispersional effects as seen on the interference figure. Figs. II–3 to II–5 are on pp. 267–269; Figs. II–6 to II–8, on pp. 282–284.

Example of the Use of Biaxial Tables

Assume that the optical constants of an unknown mineral have been determined to be $\alpha = 1.681$, $\beta = 1.706$, $\gamma = 1.718$, $(-)$. The values $(\gamma - \alpha)$ and $(\gamma - \beta)$—.037 and .012 respectively—locate a data point on Fig. II–7. Record those code numbers (that is, 308, 384, 444, 451, 453, 454, and 457) that occur within 0.002 (in all directions) of this data point. Note, however, from the inset "Range of β values" in Fig. II–7, that only code numbers 377 to 426 of Table II–4 represent minerals whose β values are near 1.706, that of the unknown. Of the code numbers recorded, only 384 falls within this group. This number is then looked up in Table II–4; it is olivine and has refractive indices identical to those of the unknown. The unknown thus appears to be another specimen of olivine; it would be wise, however, to consult Winchell and Winchell (1951, p. 498) to obtain a fuller description of olivine for a more complete comparison with the properties of the unknown.

Table II–3

LIST OF BIAXIAL POSITIVE MINERALS

(arranged in order of increasing values of β)

Code number	β	α	γ	Name
1.	1.413	1.411	1.420	Pachnolite
2.	1.44	1.438	1.452	Erionite
3.	1.441	1.439	1.469	Stercorite
4.	1.448	1.447	1.459	Taylorite
5.	1.463	1.461	1.476	Picromerite
6.	1.464	1.459	1.470	Aluminite
7.	1.468	1.456	1.507	Lansfordite
8.	1.470	1.469	1.479	Boussingaultite
9.	1.47	1.469	1.473	Tridymite
10.	1.472	1.471	1.474	Tridymite
11.	1.472	1.467	1.497	Carnallite
12.	1.473	1.468	1.481	Thenardite
13.	1.474	1.467	1.496	Carnallite
14.	1.474	1.464	1.485	Thenardite
15.	1.474	1.472	1.476	Mordenite
16.	1.475	1.467	1.494	Carnallite
17.	1.476	1.474	1.483	Alunogen
18.	1.476	1.474	1.478	Arduinite
19.	1.476	1.473	1.485	Natrolite
20.	1.477	1.471	1.484	Thenardite
21.	1.478	1.471	1.486	Melanterite
22.	1.479	1.478	1.481	Tridymite
23.	1.479	1.478	1.482	Ferrierite
24.	1.479	1.472	1.487	Pisanite
25.	1.48	1.47	1.49	Boothite
26.	1.480	1.479	1.483	Tridymite
27.	1.480	1.475	1.487	Misenite
28.	1.480	1.475	1.488	Dietrichite
29.	1.482	1.480	1.493	Natrolite
30.	1.482	1.481	1.486	Ashtonite
31.	1.483	1.479	1.488	Melanterite, Zn-Cu
32.	1.485	1.485	1.488	Chabazite
33.	1.485	1.482	1.489	Heulandite
34.	1.486	1.484	1.502	Cyanochroite
35.	1.487	1.484	1.496	Tamarugite
36.	1.487	1.470	1.540	Ammonioborite
37.	1.490	1.473	1.511	Fluellite
38.	1.494	1.493	1.497	Arcanite
39.	1.495	1.489	1.506	Kreuzbergite
40.	1.496	1.495	1.504	Struvite

Table II–3 (continued)

LIST OF BIAXIAL POSITIVE MINERALS

Code number	β	α	γ	Name
41.	1.496	1.492	1.500	Dachiardite
42.	1.497	1.496	1.501	Heulandite
43.	1.499	1.498	1.505	Heulandite
44.	1.500	1.498	1.503	Phillipsite
45.	1.50	1.498	1.503	Wellsite
46.	1.503	1.499	1.538	Uranothallite
47.	1.503	1.501	1.510	Prosopite
48.	1.504	1.493	1.517	Chrysotile
49.	1.504	1.491	1.520	Ulexite
50.	1.505	1.503	1.508	Harmotome
51.	1.505	1.496	1.519	Ulexite
52.	1.505	1.505	1.506	Mesolite
53.	1.507	1.494	1.528	Bischofite
54.	1.507	1.507	1.508	Mesolite
55.	1.508	1.504	1.545	Ussingite
56.	1.509	1.493	1.561	Larderellite
57.	1.510	1.504	1.575	Pirssonite
58.	1.510	1.504	1.516	Petalite
59.	1.512	1.510	1.523	Brewsterite
60.	1.512	1.508	1.522	Chrysotile
61.	1.512	1.505	1.524	Flagstaffite
62.	1.513	1.511	1.518	Thomsonite
63.	1.513	1.512	1.518	Faroelite
64.	1.518	1.514	1.533	Newberyite
65.	1.518	1.516	1.533	Felsobanyite
66.	1.520	1.510	1.543	Bobierite
67.	1.523	1.520	1.530	Gypsum
68.	1.523	1.521	1.530	Gypsum
69.	1.523	1.521	1.533	Mascagnite
70.	1.523	1.518	1.588	Hatchettite
71.	1.525	1.515	1.544	Probertite
72.	1.525	1.520	1.540	Thomsonite
73.	1.525	1.521	1.545	Spadaite
74.	1.525	1.523	1.532	Chalcoalumite
75.	1.526	1.508	1.550	Kaliborite
76.	1.527	1.523	1.545	Hydromagnesite
77.	1.528	1.509	1.548	Paternoite
78.	1.529	1.507	1.573	Copiapite
79.	1.529	1.525	1.536	Albite
80.	1.529	1.522	1.577	Botryogen

Table II–3 (continued)

LIST OF BIAXIAL POSITIVE MINERALS

Code number	β	α	γ	Name
81.	1.530	1.522	1.544	Tavistockite
82.	1.532	1.529	1.539	Albite
83.	1.533	1.520	1.584	Kieserite
84.	1.534	1.533	1.575	Fibroferrite
85.	1.534	1.523	1.570	Hydroboracite
86.	1.534	1.525	1.552	Wavellite
87.	1.540	1.536	1.544	Albiclase
88.	1.543	1.542	1.555	Chrysotile
89.	1.543	1.534	1.588	Gordonite
90.	1.543	1.535	1.561	Wavellite
91.	1.543	1.530	1.595	Copiapite
92.	1.544	1.544	1.546	Epididymite
93.	1.545	1.539	1.551	Brushite
94.	1.546	1.545	1.551	Eudidymite
95.	1.547	1.541	1.564	Voglite
96.	1.550	1.546	1.557	Chrysotile
97.	1.550	1.530	1.597	Copiapite
98.	1.553	1.550	1.557	Andesine
99.	1.555	1.546	1.587	Bombiccite
100.	1.555	1.551	1.562	Vauxite
101.	1.556	1.553	1.560	Andesine
102.	1.558	1.554	1.573	Paravauxite
103.	1.558	1.544	1.581	Louderbackite
104.	1.558	1.551	1.582	Metavariscite
105.	1.559	1.555	1.563	Anemousite
106.	1.560	1.557	1.565	Labradorite
107.	1.561	1.550	1.577	Metavauxite
108.	1.562	1.560	1.566	Dickite
109.	1.563	1.561	1.567	Dickite
110.	1.563	1.559	1.568	Labradorite
111.	1.565	1.560	1.574	Elpidite
112.	1.565	1.561	1.570	Labratownite
113.	1.566	1.560	1.581	Humite
114.	1.566	1.566	1.587	Gibbsite
115.	1.566	1.547	1.594	Quenstedtite
116.	1.567	1.563	1.590	Norbergite
117.	1.568	1.565	1.580	Isoclasite
118.	1.57	1.555	1.585	Tengerite
119.	1.570	1.570	1.586	Gibbsite
120.	1.570	1.569	1.582	Wagnerite

Table II–3 (continued)

List of Biaxial Positive Minerals

Code number	β	α	γ	Name
121.	1.571	1.571	1.576	Clinochlore
122.	1.571	1.563	1.596	Hoernesite
123.	1.575	1.575	1.590	Leuchtenbergite
124.	1.575	1.570	1.614	Anhydrite
125.	1.576	1.570	1.614	Anhydrite
126.	1.576	1.576	1.579	Penninite
127.	1.576	1.574	1.588	Augelite
128.	1.579	1.576	1.597	Cookeite
129.	1.579	1.579	1.584	Clinochlore
130.	1.580	1.578	1.586	Sheridanite
131.	1.581	1.567	1.638	Kornelite
132.	1.581	1.578	1.588	Chlorite
133.	1.583	1.583	1.595	Zonotlite
134.	1.585	1.583	1.590	Bavenite
135.	1.585	1.578	1.591	Anthophyllite
136.	1.586	1.586	1.595	Wardite
137.	1.589	1.588	1.599	Prochlorite
138.	1.589	1.584	1.594	Celsian
139.	1.590	1.587	1.597	Custerite
140.	1.590	1.587	1.597	Bulfonteinite
141.	1.591	1.560	1.631	Hambergite
142.	1.592	1.591	1.627	Catapleiite
143.	1.592	1.586	1.614	Colemanite
144.	1.595	1.590	1.602	Cuspidine
145.	1.595	1.562	1.632	Szmikite
146.	1.596	1.577	1.616	Johannite
147.	1.597	1.597	1.612	Amesite
148.	1.598	1.580	1.627	Vivianite
149.	1.600	1.570	1.636	Vivianite
150.	1.60(5)	1.595	1.628	Cebollite
151.	1.602	1.590	1.638	Haidingerite
152.	1.603	1.594	1.615	Fremontite
153.	1.603	1.579	1.633	Vivianite
154.	1.604	1.595	1.633	Pectolite
155.	1.605	1.600	1.636	Pectolite
156.	1.605	1.600	1.613	Hydrophilite
157.	1.606	1.599	1.621	Scawtite
158.	1.606	1.606	1.610	Prochlorite
159.	1.607	1.605	1.613	Corundophilite
160.	1.608	1.600	1.645	Weinschenkite

Table II–3 (continued)

LIST OF BIAXIAL POSITIVE MINERALS

Code number	β	α	γ	Name
161.	1.610	1.607	1.618	Topaz
162.	1.610	1.604	1.636	Pectolite
163.	1.612	1.612	1.616	Aphrosiderite
164.	1.613	1.602	1.649	Anapaite
165.	1.613	1.609	1.619	Stokesite
166.	1.613	1.606	1.623	Soda tremolite
167.	1.613	1.603	1.626	Amblygonite
168.	1.614	1.607	1.630	Montebrasite
169.	1.614	1.600	1.631	Monetite
170.	1.614	1.610	1.642	Pectolite
171.	1.615	1.612	1.624	Zippeite
172.	1.616	1.613	1.623	Topaz
173.	1.617	1.614	1.636	Calamine
174.	1.617	1.588	1.655	Cyanotrichite
175.	1.617	1.604	1.636	Chondrodite
176.	1.618	1.616	1.625	Topaz
177.	1.618	1.618	1.621	Ripidolite
178.	1.618	1.612	1.642	Prehnite
179.	1.619	1.614	1.633	Edenite
180.	1.62	1.61	1.65	Turquois
181.	1.620	1.619	1.627	Topaz
182.	1.620	1.620	1.654	Churchite
183.	1.620	1.617	1.634	Afwillite
184.	1.621	1.614	1.641	Prehnite
185.	1.621	1.616	1.635	Pargasite
186.	1.621	1.620	1.630	Zippeite
187.	1.622	1.618	1.658	Arakawaite
188.	1.623	1.613	1.636	Amblygonite
189.	1.623	1.613	1.643	Chondrodite
190.	1.623	1.622	1.631	Celestite
191.	1.624	1.622	1.642	Scorodite
192.	1.624	1.622	1.631	Celestite
193.	1.624	1.615	1.644	Prehnite
194.	1.625	1.615	1.645	Prehnite
195.	1.625	1.615	1.665	Destinezite
196.	1.625	1.614	1.637	Parahopeite
197.	1.626	1.616	1.649	Prehnite
198.	1.629	1.626	1.652	Prehnite
199.	1.630	1.630	1.635	Chlorite
200.	1.630	1.622	1.645	Edenite

Table II–3 (continued)

List of Biaxial Positive Minerals

Code number	β	α	γ	Name
201.	1.630	1.619	1.650	Chondrodite
202.	1.630	1.623	1.684	Guildite
203.	1.631	1.629	1.638	Topaz
204.	1.632	1.622	1.652	Humite
205.	1.632	1.631	1.640	Picropharmacolite
206.	1.633	1.633	1.635	Ripidolite
207.	1.634	1.627	1.646	Cummingtonite
208.	1.634	1.623	1.655	Humite
209.	1.635	1.617	1.652	Tilleyite
210.	1.635	1.613	1.657	Sklodowskite
211.	1.636	1.631	1.660	Schizolite
212.	1.636	1.623	1.651	Clinohumite
213.	1.637	1.636	1.648	Barite
214.	1.638	1.633	1.652	Anthophyllite
215.	1.638	1.625	1.653	Clinohumite
216.	1.638	1.633	1.652	Pargasite
217.	1.640	1.640	1.657	Sarcolite
218.	1.642	1.638	1.653	Mullite
219.	1.642	1.632	1.657	Collinsite
220.	1.642	1.632	1.665	Ferriprehnite
221.	1.642	1.632	1.665	Prehnite
222.	1.642	1.640	1.647	Juanite
223.	1.643	1.631	1.695	Ransomite
224.	1.644	1.641	1.650	Rankinite
225.	1.644	1.632	1.664	Clinohumite
226.	1.644	1.636	1.654	Fairfieldite
227.	1.644	1.642	1.654	Mullite
228.	1.645	1.635	1.663	Chondrodite
229.	1.645	1.643	1.651	Rinkolite
230.	1.648	1.637	1.676	Loseyite
231.	1.649	1.646	1.658	Mosandrite
232.	1.650	1.639	1.667	Cummingtonite
233.	1.650	1.640	1.660	Fairfieldite
234.	1.651	1.636	1.669	Forsterite
235.	1.651	1.635	1.670	Forsterite
236.	1.653	1.650	1.658	Enstatite
237.	1.653	1.643	1.675	Humite
238.	1.653	1.640	1.680	Messelite
239.	1.653	1.651	1.665	Triplite
240.	1.654	1.651	1.660	Clinoenstatite

Table II–3 (continued)

List of Biaxial Positive Minerals

Code number	β	α	γ	Name
241.	1.655	1.655	1.662	Uranochalcite
242.	1.655	1.648	1.662	Dickinsonite, spodumene
243.	1.655	1.643	1.670	Chondrodite
244.	1.655	1.642	1.661	Anthophyllite
245.	1.656	1.653	1.673	Euclase
246.	1.656	1.649	1.714	Natrochalcite
247.	1.656	1.651	1.683	Reddingite
248.	1.658	1.652	1.665	Hiortdahlite
249.	1.658	1.655	1.667	Spodumene
250.	1.658	1.640	1.695	Veszelyite
251.	1.659	1.653	1.677	Spodumene
252.	1.659	1.654	1.667	Jadeite
253.	1.660	1.656	1.672	Spodumene
254.	1.660	1.650	1.672	Triplite
255.	1.660	1.657	1.667	Enstatite
256.	1.660	1.650	1.680	Barrandite
257.	1.660	1.645	1.715	Plancheite
258.	1.660	1.659	1.680	Sillimanite
259.	1.661	1.645	1.688	Leucosphenite
260.	1.661	1.640	1.680	Forsterite
261.	1.662	1.658	1.671	Dickinsonite
262.	1.662	1.658	1.668	Boracite
263.	1.662	1.655	1.683	Reddingite
264.	1.664	1.660	1.688	Sérandite
265.	1.666	1.661	1.673	Johnstrupite
266.	1.666	1.663	1.673	Lithiophilite
267.	1.666	1.660	1.676	Spodumene
268.	1.667	1.644	1.716	Plancheite
269.	1.667	1.653	1.678	Spodumene
270.	1.667	1.662	1.673	Boracite
271.	1.668	1.665	1.681	Rinkite
272.	1.669	1.655	1.670	Cummingtonite
273.	1.669	1.665	1.674	Enstatite
274.	1.670	1.658	1.690	Titanohydroclinohumite
275.	1.670	1.661	1.684	Sillimanite
276.	1.670	1.653	1.689	Olivine
277.	1.671	1.662	1.691	Viridine
278.	1.671	1.670	1.689	Hinsdalite
279.	1.671	1.664	1.694	Diopside
280.	1.672	1.669	1.677	Magnesium chlorophoenicite

Table II–3 (continued)
LIST OF BIAXIAL POSITIVE MINERALS

Code number	β	α	γ	Name
281.	1.672	1.672	1.676	Fillowite
282.	1.673	1.665	1.682	Triplite
283.	1.673	1.664	1.698	Clinohumite
284.	1.674	1.671	1.684	Natrophilite
285.	1.674	1.663	1.699	Spodiosite
286.	1.674	1.665	1.684	Lawsonite
287.	1.674	1.666	1.688	Diopside-jadeite
288.	1.675	1.666	1.688	Iron anthophyllite
289.	1.675	1.653	1.697	Ludlamite
290.	1.676	1.672	1.683	Akrochordite
291.	1.678	1.669	1.700	Iron reddingite
292.	1.678	1.655	1.700	Clinohumite
293.	1.678	1.673	1.683	Hypersthene
294.	1.679	1.676	1.687	Lithiophilite
295.	1.680	1.679	1.692	Pumpellyite
296.	1.681	1.663	1.700	Olivine
297.	1.683	1.675	1.692	Triplite
298.	1.683	1.662	1.717	Koettigite
299.	1.683	1.676	1.705	Zinc schefferite
300.	1.685	1.678	1.703	Leucaugite
301.	1.686	1.681	1.698	Titanoelpidite
302.	1.687	1.686	1.693	Triphylite
303.	1.687	1.682	1.711	Rosenbuschite
304.	1.688	1.679	1.710	Urbanite
305.	1.688	1.688	1.692	Triphylite
306.	1.688	1.685	1.698	Thulite
307.	1.690	1.682	1.698	Chlorophoenicite
308.	1.690	1.670	1.706	Cummingtonite
309.	1.690	1.682	1.710	Jeffersonite
310.	1.691	1.690	1.711	Pigeonite
311.	1.692	1.687	1.709	Fowlerite
312.	1.694	1.687	1.713	Augite
313.	1.695	1.694	1.702	Triphylite
314.	1.696	1.691	1.703	Barylite
315.	1.696	1.696	1.702	Zoisite
316.	1.698	1.695	1.733	Euchroite
317.	1.699	1.690	1.736	Neptunite
318.	1.699	1.690	1.721	Schefferite
319.	1.699	1.692	1.721	Salite
320.	1.700	1.691	1.724	Clinohumite

Table II–3 (*continued*)

LIST OF BIAXIAL POSITIVE MINERALS

Code number	β	α	γ	Name
321.	1.700	1.697	1.703	Riebeckite
322.	1.70	1.695	1.722	Gadolinite
323.	1.702	1.696	1.714	Augite
324.	1.702	1.700	1.706	Zoisite
325.	1.703	1.701	1.706	Serendibite
326.	1.703	1.700	1.718	Zoisite
327.	1.704	1.696	1.713	Triplite
328.	1.705	1.700	1.724	Graftonite
329.	1.705	1.691	1.738	Astrophyllite
330.	1.705	1.691	1.735	Astrophyllite
331.	1.706	1.700	1.724	Augite
332.	1.707	1.700	1.718	Pumpellyite
333.	1.708	1.702	1.726	Diopside-hedenbergite
334.	1.708	1.708	1.745	Strengite
335.	1.709	1.702	1.741	Legrandite
336.	1.710	1.705	1.725	Thulite
337.	1.711	1.708	1.718	Merwinite
338.	1.711	1.709	1.724	Brandtite
339.	1.712	1.706	1.724	Merwinite
340.	1.713	1.713	1.738	Pigeonite
341.	1.713	1.703	1.722	Gerhardtite
342.	1.714	1.704	1.735	Ferrosalite
343.	1.714	1.714	1.744	Pigeonite
344.	1.715	1.707	1.730	Larnite
345.	1.715	1.711	1.724	Rhodonite
346.	1.717	1.715	1.719	Clinozoisite
347.	1.718	1.715	1.733	Magnesium orthite
348.	1.718	1.699	1.742	Augite
349.	1.719	1.717	1.741	Pigeonite
350.	1.719	1.710	1.738	Johannsenite
351.	1.719	1.715	1.737	Chloritoid
352.	1.720	1.710	1.736	Hedenbergite, Mn
353.	1.720	1.715	1.725	Clinozoisite
354.	1.720	1.718	1.723	Clinozoisite
355.	1.720	1.716	1.728	Rhodonite
356.	1.721	1.712	1.731	Adelite
357.	1.722	1.702	1.750	Diaspore
358.	1.722	1.713	1.745	Jeffersonite
359.	1.722	1.720	1.731	Chloritoid
360.	1.725	1.722	1.728	Chloritoid

Table II–3 (continued)
List of Biaxial Positive Minerals

Code number	β	α	γ	Name
361.	1.725	1.715	1.738	Homilite
362.	1.725	1.721	1.746	Titanaugite
363.	1.726	1.724	1.737	Chloritoid
364.	1.726	1.725	1.730	Triploidite
365.	1.728	1.725	1.737	Iron rhodonite
366.	1.730	1.726	1.737	Fowlerite
367.	1.730	1.717	1.752	Babingtonite
368.	1.730	1.718	1.751	Roselite
369.	1.730	1.725	1.746	Augite, Ti
370.	1.732	1.726	1.753	Ferroaugite
371.	1.732	1.730	1.762	Strengite
372.	1.732	1.726	1.751	Hedenbergite
373.	1.735	1.726	1.752	Hedenbergite
374.	1.737	1.733	1.747	Rhodonite
375.	1.738	1.726	1.789	Antlerite
376.	1.740	1.739	1.760	Ardennite
377.	1.740	1.732	1.757	Hedenbergite
378.	1.741	1.736	1.746	Staurolite
379.	1.742	1.738	1.765	Scorodite
380.	1.744	1.739	1.750	Staurolite
381.	1.745	1.739	1.751	Staurolite
382.	1.746	1.745	1.765	Astrophyllite
383.	1.747	1.744	1.773	Molengraaffite
384.	1.748	1.747	1.757	Chrysoberyl
385.	1.750	1.744	1.756	Staurolite
386.	1.754	1.747	1.762	Staurolite
387.	1.754	1.745	1.780	Lamprophyllite
388.	1.758	1.730	1.838	Azurite
389.	1.763	1.762	1.765	Dewindtite
390.	1.764	1.746	1.806	Piedmontite
391.	1.765	1.739	1.799	Piedmontite
392.	1.770	1.769	1.785	Holdenite
393.	1.772	1.754	1.795	Piedmontite
394.	1.774	1.765	1.797	Scorodite (?)
395.	1.774	1.770	1.783	Barthite
396.	1.776	1.758	1.795	Orientite
397.	1.780	1.776	1.805	Caryinite
398.	1.782	1.752	1.815	Shattuckite
399.	1.786	1.775	1.815	Beraunite
400.	1.787	1.785	1.840	Monazite

Table II–3 (*continued*)

LIST OF BIAXIAL POSITIVE MINERALS

Code number	β	α	γ	Name
401.	1.788	1.783	1.818	Lossenite (= Beudantite)
402.	1.788	1.777	1.800	Retzian
403.	1.788	1.786	1.837	Monazite
404.	1.789	1.756	1.829	Piedmontite
405.	1.794	1.794	1.803	Arseniopleite
406.	1.795	1.780	1.815	Barthite
407.	1.796	1.784	1.814	Scorodite
408.	1.801	1.800	1.849	Monazite
409.	1.801	1.783	1.834	Flinkite
410.	1.805	1.78	1.82	Tephroite
411.	1.810	1.772	1.863	Olivenite
412.	1.810	1.808	1.830	Warwickite
413.	1.812	1.801	1.824	Gadolinite
414.	1.818	1.817	1.821	Cerite
415.	1.820	1.780	1.865	Olivenite
416.	1.827	1.792	1.864	Iddingsite
417.	1.840	1.830	1.885	Dufrenite
418.	1.852	1.845	1.878	Toernebohmite
419.	1.870	1.843	1.943	Sphene
420.	1.880	1.873	1.895	Rockbridgeite
421.	1.882	1.877	1.894	Anglesite
422.	1.883	1.878	1.895	Anglesite
424.	1.910	1.895	1.950	Kasolite
423.	1.920	1.885	1.956	Tsumebite
425.	1.92	1.87	2.01	Claudetite
426.	1.963	1.963	1.966	Hyalotekite

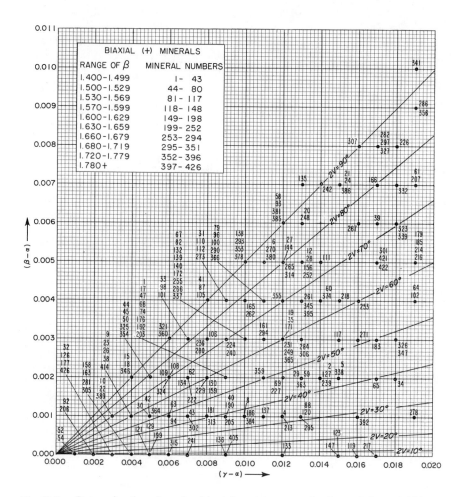

Fig. II-3. Determinative chart for biaxial positive crystals, $(\gamma - \alpha)$ from 0.001 to 0.019. Consult Table II-3 for identity of code numbers used in chart.

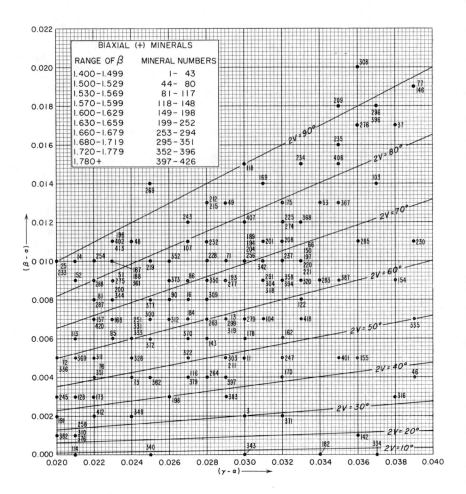

Fig. II-4. Determinative chart for biaxial positive crystals, $(\gamma - \alpha)$ from 0.020 to 0.039. Consult Table II-3 for identity of code numbers used in chart.

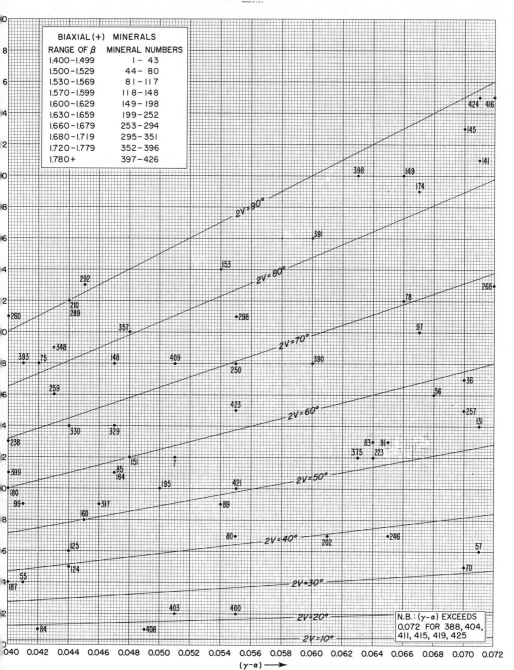

Fig. II-5. Determinative chart for biaxial positive crystals, $(\gamma - \alpha)$ from 0.040 to 0.072. Consult Table II-3 for identity of code numbers used in chart.

Table II–4

LIST OF BIAXIAL NEGATIVE MINERALS

(arranged in order of increasing values of β)

Code number	β	α	γ	Name
1.	1.3245	1.3239	1.3247	Avogadrite
2.	1.395	1.393	1.397	Mirabilite
3.	1.396	1.394	1.398	Mirabilite
4.	1.414	1.407	1.415	Thomsenolite
5.	1.425	1.405	1.440	Natron
6.	1.452	1.440	1.453	Lecontite
7.	1.452	1.430	1.458	Kalinite
8.	1.453	1.426	1.456	Hexahydrite
9.	1.454	1.448	1.456	Gearksutite
10.	1.455	1.435	1.459	Wattevillite
11.	1.455	1.433	1.461	Epsomite
12.	1.456	1.340	1.459	Sassolite
13.	1.457	1.432	1.458	Mendozite
14.	1.461	1.449	1.463	Mendozite
15.	1.469	1.447	1.472	Borax
16.	1.470	1.462	1.471	Paraluminite
17.	1.470	1.447	1.473	Goslarite
18.	1.470	1.447	1.472	Borax
19.	1.472	1.454	1.488	Kernite
20.	1.475	1.474	1.476	Ptilolite
21.	1.475	1.472	1.476	Mordenite
22.	1.478	1.461	1.485	Creedite
23.	1.479	1.476	1.479	Clinoptilolite
24.	1.480	1.457	1.484	Goslarite
25.	1.480	1.476	1.483	Pickeringite
26.	1.48	1.47	1.49	Boothite
27.	1.482	1.478	1.482	Apjohnite
28.	1.483	1.477	1.489	Bieberite
29.	1.486	1.483	1.487	Bloedite
30.	1 486	1.480	1.490	Halotrichite
31.	1.487	1.483	1.490	Leonite
32.	1.488	1.485	1.489	Vanthoffite
33.	1.488	1.480	1.490	Halotrichite
34.	1.489	1.467	1.492	Morenosite
35.	1.489	1.469	1.492	Morenosite
36.	1.491	1.482	1.493	Stilbite
37.	1.492	1.484	1.495	Stilbite, stellerite
38.	1.492	1.412	1.540	Trona
39.	1.494	1.465	1.495	Bianchite
40.	1.497	1.493	1.500	Phillipsite

Table II–4 (continued)
List of Biaxial Negative Minerals

Code number	β	α	γ	Name
41.	1.498	1.465	1.504	Nitrocalcite
42.	1.498	1.489	1.499	Stilbite
43.	1.498	1.487	1.500	Stilbite
44.	1.498	1.494	1.500	Stilbite
45.	1.499	1.492	1.501	Stilbite
46.	1.500	1.490	1.502	Stilbite
47.	1.501	1.493	1.507	Stilbite, epidesmine
48.	1.501	1.493	1.508	Didymolite
49.	1.502	1.490	1.511	Antigorite
50.	1.505	1.494	1.516	Kainite
51.	1.505	1.492	1.516	Inyoite
52.	1.506	1.498	1.508	Parasepiolite, Mn
53.	1.508	1.495	1.514	Chalcanthite
54.	1.51	1.495	1.520	Inyoite
55.	1.510	1.502	1.512	Epistilbite
56.	1.510	1.479	1.511	Saponite
57.	1.510	1.49	1.521	Uranospathite
58.	1.512	1.506	1.517	Leonhardite
59.	1.514	1.509	1.514	Carnegieite
60.	1.514	1.504	1.516	Laumontite
61.	1.514	1.512	1.515	Okenite
62.	1.515	1.505	1.517	Laumontite
63.	1.516	1.493	1.516	Montmorillonite
64.	1.517	1.500	1.518	Syngenite
65.	1.518	1.511	1.522	Laumontite
66.	1.518	1.508	1.522	Ilesite
67.	1.519	1.512	1.519	Scolecite
68.	1.526(?)	1.519	1.529	Sepiolite
69.	1.52	1.516	1.520	Carnegieite
70.	1.522	1.518	1.525	Microcline
71.	1.524	1.513	1.525	Laumontite
72.	1.524	1.518	1.526	Orthoclase
73.	1.525	1.519	1.527	Anorthoclase
74.	1.526	1.503	1.527	Montmorillonite
75.	1.526	1.522	1.530	Microcline
76.	1.527	1.490	1.527	Saponite
77.	1.528	1.523	1.529	Analbite (?)
78.	1.530	1.518	1.542	Minasragrite
79.	1.531	1.528	1.534	Hyalophane
80.	1.531	1.525	1.535	Searlesite

Table II–4 (continued)
List of Biaxial Negative Minerals

Code number	β	α	γ	Name
81.	1.533	1.513	1.535	Searlesite
82.	1.534	1.489	1.557	Artinite
83.	1.534	1.514	1.541	Chalcanthite, Zn-Cu
84.	1.535	1.500	1.560	Meyerhofferite
85.	1.535	1.515	1.536	Glauberite
86.	1.535	1.530	1.540	Okenite
87.	1.535	1.494	1.536	Beidellite
88.	1.536	1.517	1.543	Chalcanthite, Fe-Cu
89.	1.537	1.495	1.537	Beidellite
90.	1.537	1.528	1.543	Siderotil
91.	1.537	1.514	1.543	Chalcanthite
92.	1.538	1.534	1.540	Hyalophane
93.	1.538	1.534	1.540	Cordierite
94.	1.540	1.527	1.544	Sulphoborite
95.	1.540	1.531	1.548	Gismondite
96.	1.541	1.520	1.545	Lueneburgite
97.	1.543	1.538	1.545	Cordierite
98.	1.543	1.539	1.547	Oligoclase (Ab₄An)
99.	1.545	1.503	1.545	Pholidolite
100.	1.545	1.542	1.547	Hyalophane
101.	1.545	1.525	1.545	Vermiculite
102.	1.545	1.533	1.547	Mooreite
103.	1.546	1.539	1.551	Brushite
104.	1.546	1.542	1.549	Oligoclase
105.	1.547	1.520	1.572	Copiapite
106.	1.547	1.540	1.550	Cordierite
107.	1.548	1.535	1.549	Centrallasite
108.	1.549	1.528	1.549	Truscottite
109.	1.549	1.531	1.552	Chalcanthite, Co
110.	1.549	1.517	1.549	Montmorillonite
111.	1.549	1.538	1.554	Edingtonite
112.	1.551	1.547	1.554	Andeclase
113.	1.553	1.530	1.556	Lepidolite
114.	1.554	1.535	1.555	Lepidolite
115.	1.554	1.545	1.565	Lacroixite
116.	1.555	1.532	1.555	Lepidolite
117.	1.556	1.552	1.559	Miloschite
118.	1.555	1.543	1.558	Lepidolite, polylithionite
119.	1.558	1.552	1.561	Beryllonite
120.	1.558	1.520	1.558	Phlogopite

Table II–4 (continued)

LIST OF BIAXIAL NEGATIVE MINERALS

Code number	β	α	γ	Name
121.	1.560	1.547	1.567	Polyhalite
122.	1.560	1.54	1.560	Vermiculite
123.	1.560	1.54	1.560	Jefferisite
124.	1.561	1.551	1.562	Jezekite
125.	1.562	1.551	1.563	Morinite
126.	1.562	1.552	1.563	Cordierite
127.	1.562	1.557	1.563	Nacrite
128.	1.563	1.560	1.566	Nacrite
129.	1.564	1.559	1.565	Anauxite
130.	1.564	1.535	1.565	Phlogopite
131.	1.565	1.543	1.565	Potash clay
132.	1.565	1.544	1.566	Lepidolite, polylithionite
133.	1.565	1.550	1.570	Variscite
134.	1.565	1.561	1.566	Kaolinite
135.	1.570	1.560	1.571	Antigorite
136.	1.570	1.565	1.574	Bytownite (AbAn$_3$)
137.	1.570	1.523	1.572	Beidellite
138.	1.571	1.524	1.583	Romerite
139.	1.571	1.554	1.576	Variscite
140.	1.572	1.555	1.575	Hannayite
141.	1.572	1.566	1.576	Bytownite (AbAn$_4$)
142.	1.572	1.570	1.572	Englishite
143.	1.573	1.542	1.573	Vermiculite
144.	1.574	1.540	1.574	Phlogopite
145.	1.574	1.56	1.580	Bassetite
146.	1.574	1.568	1.578	Bytownorthite (Ab$_3$An$_{17}$)
147.	1.575	1.553	1.577	Autunite
148.	1.576	1.546	1.576	Stilpnomelane
149.	1.576	1.562	1.588	Sphaerite
150.	1.577	1.565	1.578	Eastonite
151.	1.578	1.576	1.579	Penninite
152.	1.578	1.551	1.581	Zinnwaldite
153.	1.578	1.544	1.601	Kroehnkite
154.	1.58	1.55	1.58	Zinnwaldite
155.	1.580	1.562	1.582	Canbyite
156.	1.580	1.573	1.585	Anorthite (AbAn$_{19}$)
157.	1.581	1.561	1.581	Vermiculite
158.	1.581	1.551	1.587	Muscovite
159.	1.581	1.559	1.586	Glauconite
160.	1.582	1.570	1.587	Celsian

Table II–4 (continued)

LIST OF BIAXIAL NEGATIVE MINERALS

Code number	β	α	γ	Name
161.	1.582	1.560	1.587	Uranospinite
167.	1.582	1.574	1.582	Beta hopeite
163.	1.582	1.552	1.588	Muscovite
164.	1.583	1.562	1.590	Variscite
165.	1.584	1.566	1.593	Variscite
166.	1.584	1.576	1.588	Anorthite
167.	1.584	1.579	1.585	Delta mooreite
168.	1.585	1.56	1.585	Nontronite
169.	1.588	1.552	1.600	Pyrophyllite
170.	1.588	1.559	1.588	Montmorillonite
171.	1.589	1.539	1.589	Talc
172.	1.589	1.583	1.594	Pharmacolite
173.	1.590	1.578	1.599	Variscite
174.	1.590	1.575	1.605	Catapleiite
175.	1.591	1.572	1.59	Alpha hopeite
176.	1.591	1.572	1.594	Priceite
177.	1.592	1.582	1.592	Torbernite
178.	1.593	1.585	1.608	Nontronite
179.	1.594	1.55	1.594	Alurgite
180.	1.594	1.570	1.597	Astrolite
181.	1.594	1.584	1.594	Clinochlore
182.	1.594	1.551	1.594	Stilpnomelane
183.	1.595	1.578	1.598	Amblygonite
184.	1.595	1.572	1.614	Johannite
185.	1.595	1.555	1.595	Diabantite
186.	1.595	1.558	1.601	Muscovite
187.	1.595	1.575	1.610	Variscite
188.	1.595	1.571	1.598	Leucophanite
189.	1.596	1.551	1.596	Fluorbiotite
190.	1.597	1.592	1.599	Cordierite
191.	1.597	1.575	1.598	Chryscolla
192.	1.598	1.568	1.598	Manganophyllite
193.	1.598	1.584	1.602	Millisite
194.	1.598	1.586	1.605	Howlite
195.	1.599	1.561	1.599	Stilpnomelane
196.	1.599	1.556	1.602	Muscovite
197.	1.599	1.564	1.600	Paragonite
198.	1.600	1.589	1.610	Nontronite
199.	1.602	1.586	1.606	Spencerite
200.	1.603	1.597	1.605	Foshagite

Table II–4 (*continued*)

LIST OF BIAXIAL NEGATIVE MINERALS

Code number	β	α	γ	Name
201.	1.603	1.573	1.604	Ganophyllite
202.	1.605	1.591	1.614	Bertrandite
203.	1.605	1.577	1.609	Paragonite
204.	1.606	1.592	1.617	Chondrodite
205.	1.606	1.562	1.606	Phlogopite
206.	1.607	1.594	1.619	Chondrodite
207.	1.61	1.605	1.612	Hillebrandite
208.	1.61	1.57	1.61	Protolithionite
209.	1.611	1.572	1.615	Muscovite
210.	1.611	1.600	1.620	Montebrasite
211.	1.612	1.592	1.621	Herderite
212.	1.613	1.593	1.613	Meliphanite
213.	1.613	1.599	1.625	Tremolite
214.	1.614	1.594	1.616	Phosphophyllite
215.	1.614	1.590	1.615	Glauconite
216.	1.615	1.605	1.622	Soda-tremolite (richterite)
217.	1.616	1.606	1.623	Soda-tremolite
218.	1.616	1.597	1.624	Sanbornite
219.	1.616	1.600	1.627	Tremolite
220.	1.616	1.602	1.629	Lehiite
221.	1.618	1.597	1.619	Glauconite
222.	1.619	1.605	1.619	Delessite
223.	1.620	1.573	1.620	Biotite
224.	1.622	1.609	1.622	Glauconite
225.	1.623	1.565	1.623	Stilpnomelane
226.	1.623	1.610	1.623	Uranocircite
227.	1.623	1.614	1.624	Dehrnite
228.	1.625	1.595	1.627	Soda margarite, ephesite
229.	1.625	1.592	1.625	Protolithionite
230.	1.628	1.611	1.630	Carpholite
231.	1.628	1.598	1.654	Lausenite
232.	1.628	1.609	1.628	Glauconite
233.	1.629	1.615	1.637	Richterite
234.	1.629	1.620	1.630	Margarite
235.	1.63±	1.60	1.63	Mariposite
236.	1.630	1.585	1.630	Troegerite
237.	1.63	1.580	1.630	Biotite
238.	1.63	1.59	1.64	Roscoelite
239.	1.63	1.608	1.638	Celadonite
240.	1.630	1.603	1.630	Glauconite

Table II–4 (continued)

LIST OF BIAXIAL NEGATIVE MINERALS

Code number	β	α	γ	Name
241.	1.630	1.614	1.641	Actinolite
242.	1.630	1.619	1.640	Anthophyllite
243.	1.630	1.622	1.630	Podolite
244.	1.631	1.618	1.642	Hastingsite
245.	1.632	1.622	1.641	Pargasite
246.	1.632	1.602	1.632	Bementite
247.	1.632	1.615	1.634	Glaucophane
248.	1.632	1.620	1.634	Wollastonite
249.	1.633	1.618	1.641	Actinolite
250.	1.633	1.630	1.636	Danburite
251.	1.633	1.629	1.639	Andalusite
252.	1.634	1.612	1.643	Lazulite
253.	1.635	1.622	1.641	Richterite
254.	1.635	1.629	1.640	Anthophyllite
255.	1.636	1.622	1.636	Manganophyllite
256.	1.636	1.602	1.639	Grandidierite
257.	1.636	1.609	1.644	Inesite
258.	1.636	1.623	1.644	Gedrite
259.	1.637	1.615	1.638	Chlorite
260.	1.638	1.621	1.638	Glaucophane
261.	1.639	1.634	1.643	Andalusite
262.	1.639	1.634	1.645	Andalusite
265.	1.639	1.624	1.643	Roscherite
264.	1.639	1.633	1.642	Arfvedsonite
265.	1.641	1.637	1.646	Andalusite
266.	1.642	1.584	1.647	Serpierite
267.	1.642	1.611	1.657	Sklodowskite
268.	1.643	1.632	1.645	Margarite
269.	1.643	1.612	1.644	Glauconite
270.	1.644	1.636	1.649	Eckermannite
271.	1.645	1.625	1.654	Holmquistite
272.	1.645	1.638	1.651	Boehmite
273.	1.646	1.639	1.653	Monticellite
274.	1.646	1.630	1.649	Gastaldite
275.	1.647	1.634	1.652	Hornblende
276.	1.647	1.624	1.647	Bementite
277.	1.648	1.584	1.648	Biotite
278.	1.649	1.643	1.649	Daphnite
279.	1.650	1.637	1.660	Mangan-tremolite
280.	1.650	1.640	1.653	Wollastonite

Table II–4 (*continued*)

LIST OF BIAXIAL NEGATIVE MINERALS

Code number	β	α	γ	Name
281.	1.650	1.623	1.651	Friedelite
282.	1.65	1.625	1.655	Nontronite
283.	1.65	1.64	1.66	Crocidolite
284.	1.650	1.610	1.682	Epistolite
285.	1.650	1.608	1.655	Iddingsite
286.	1.651	1.648	1.651	Aphrosiderite
287.	1.652	1.643	1.654	Bityite
288.	1.652	1.612	1.675	Liroconite
289.	1.654	1.626	1.670	Datolite
290.	1.654	1.62	1.689	Cabrerite
291.	1.654	1.647	1.660	Hureaulite
292.	1.655	1.633	1.662	Eosphorite
293.	1.655	1.646	1.657	Wentzelite
294.	1.655	1.642	1.661	Anthophyllite
295.	1.656	1.652	1.660	Palaite
296.	1.657	1.652	1.662	Baldaufite
297.	1.657	1.644	1.663	Pargasite
298.	1.657	1.646	1.658	Seybertite
299.	1.658	1.622	1.687	Annabergite
300.	1.660	1.648	1.660	Brandisite
301.	1.660	1.640	1.675	Tilasite
302.	1.660	1.649	1.661	Xanthophyllite
303.	1.66	1.63	1.69	Stewartite
304.	1.660	1.638	1.667	Eosphorite
305.	1.660	1.650	1.660	Chlorite
306.	1.661	1.653	1.669	Hastingsite
307.	1.662	1.651	1.668	Monticellite
308.	1.663	1.637	1.673	Lazulite
309.	1.663	1.640	1.665	Seamanite
310.	1.664	1.516	1.666	Strontianite
311.	1.664	1.655	1.668	Ternovskite
312.	1.665	1.650	1.679	Cummingtonite
313.	1.665	1.653	1.665	Thuringite
314.	1.665	1.650	1.675	Hornblende
315.	1.666	1.654	1.670	Hornblende
316.	1.667	1.662	1.669	Clinohedrite
317.	1.667	1.642	1.669	Uranophane
318.	1.667	1.658	1.667	Prochlorite
319.	1.668	1.635	1.702	Symplesite
320.	1.669	1.658	1.670	Zinkosite

Table II–4 (continued)

LIST OF BIAXIAL NEGATIVE MINERALS

Code number	β	α	γ	Name
321.	1.670	1.618	1.670	Siderophyllite
322.	1.67	1.65	1.67	Strigovite
323.	1.671	1.662	1.691	Andalusite
324.	1.672	1.664	1.680	Hastingsite
325.	1.673	1.634	1.685	Durangite
326.	1.673	1.662	1.684	Triplite
327.	1.673	1.659	1.681	Hornblende, green
328.	1.674	1.640	1.679	Spurrite
329.	1.674	1.662	1.676	Bustamite
330.	1.674	1.657	1.685	Cummingtonite
331.	1.675	1.664	1.679	Bustamite
332.	1.676	1.529	1.677	Witherite
333.	1.676	1.665	1.677	Kornerupine
334.	1.676	1.623	1.677	Biotite
335.	1.677	1.663	1.685	Soretite
336.	1.678	1.643	1.684	Childrenite
337.	1.680	1.670	1.682	Arfvedsonite
338.	1.681	1.530	1.685	Aragonite
339.	1.682	1.610	1.692	Roscoelite
340.	1.683	1.670	1.693	Hornblende, basaltic
341.	1.684	1.676	1.689	Axinite
342.	1.684	1.663	1.699	Grünerite
343.	1.684	1.659	1.686	Dumortierite
344.	1.685	1.678	1.688	Axinite
345.	1.685	1.675	1.690	Dumortierite
346.	1.685	1.672	1.687	Bustamite
347.	1.685	1.595	1.685	Stilpnomelane
348.	1.685	1.658	1.690	Schroeckingerite
349.	1.685	1.670	1.685	Thuringite
350.	1.686	1.67	1.698	Trichalcite
351.	1.686	1.678	1.689	Dumortierite
352.	1.687	1.668	1.688	Dumortierite
353.	1.689	1.682	1.694	Hypersthene
354.	1.689	1.664	1.692	Cenosite
355.	1.690	1.675	1.693	Sincosite
356.	1.690	1.682	1.697	Chlorophoenicite
357.	1.690	1.630	1.690	Annite
358.	1.691	1.670	1.692	Dumortierite
359.	1.691	1.675	1.701	Hornblende
360.	1.692	1.674	1.712	Chrysolite

Table II–4 (continued)

LIST OF BIAXIAL NEGATIVE MINERALS

Code number	β	α	γ	Name
361.	1.692	1.684	1.696	Axinite
362.	1.692	1.597	1.692	Stilpnomelane
363.	1.693	1.676	1.707	Grünerite
364.	1.694	1.676	1.708	Kaersutite
365.	1.694	1.679	1.698	Hastingsite
366.	1.695	1.542	1.699	Aragonite
367.	1.695	1.675	1.714	Forsterite
368.	1.695	1.680	1.698	Tschermakite
369.	1.695	1.691	1.696	Triphylite
370.	1.695	1.684	1.698	Kempite
371.	1.697	1.672	1.717	Grünerite
372.	1.698	1.680	1.700	Hornblende
373.	1.698	1.650	1.711	Oxyhornblende
374.	1.700	1.696	1.702	Triphylite
375.	1.700	1.685	1.703	Willemite
276.	1.701	1.687	1.703	Bustamite
377.	1.701	1.680	1.720	Olivine
378.	1.701	1.695	1.703	Triphylite
379.	1.701	1.693	1.704	Tinzenite
380.	1.702	1.692	1.705	Hypersthene
381.	1.702	1.695	1.708	Barylite
382.	1.705	1.660	1.713	Tarbuttite
383.	1.705	1.692	1.707	Bustamite
384.	1.706	1.681	1.718	Olivine
385.	1.707	1.687	1.708	Barkevikite
386.	1.707	1.704	1.710	Sapphire
387.	1.709	1.686	1.729	Grünerite
388.	1.710	1.695	1.710	Hastingsite
389.	1.713	1.703	1.722	Gerhardtite
390.	1.714	1.697	1.722	Strengite
391.	1.714	1.690	1.735	Schoepite
392.	1.715	1.675	1.735	Oxyhornblende
393.	1.715	1.674	1.718	Iddingsite
394.	1.716	1.679	1.729	Glaucochroite
395.	1.716	1.700	1.726	Woehlerite
396.	1.719	1.716	1.723	Epidote
397.	1.719	1.698	1.722	Ferrohastingsite
398.	1.719	1.699	1.721	Hastingsite
399.	1.719	1.715	1.720	Vesuvianite
400.	1.720	1.715	1.725	Trimerite

Table II–4 (continued)

LIST OF BIAXIAL NEGATIVE MINERALS

Code number	β	α	γ	Name
401.	1.720	1.712	1.728	Kyanite
402.	1.720	1.691	1.720	Phosphuranylite
403.	1.722	1.687	1.731	Tarapacaite
404.	1.722	1.717	1.729	Kyanite
405.	1.722	1.713	1.729	Kyanite
406.	1.723	1.698	1.745	Lavenite
407.	1.725	1.692	1.738	Phosphosiderite
408.	1.725	1.680	1.752	Hornblende, basaltic
409.	1.726	1.694	1.730	Tyrolite
410.	1.727	1.711	1.740	Picrotephroite
411.	1.728	1.715	1.731	Hypersthene
412.	1.728	1.675	1.730	Sarcopside
413.	1.728	1.720	1.735	Landesite
414.	1.730	1.692	1.760	Kearsutite
415.	1.731	1.710	1.732	Chalcomenite
416.	1.733	1.724	1.739	Hydrocyanite
417.	1.734	1.723	1.736	Gageite
418.	1.734	1.696	1.743	Monticellite
419.	1.735	1.625	1.735	Stilpnomelane
420.	1.735	1.715	1.745	Sicklerite
421.	1.735	1.712	1.753	Hyalosiderite
422.	1.736	1.715	1.739	Renardite
423.	1.738	1.731	1.744	Thalenite
424.	1.739	1.727	1.751	Allanite, altered
425.	1.740	1.733	1.744	Rhodonite
426.	1.742	1.722	1.750	Epidote
427.	1.742	1.724	1.746	Hodgkinsonite
428.	1.744	1.708	1.773	Adamite
429.	1.745	1.723	1.765	Iddingsite
430.	1.749	1.724	1.749	Allodelphite
431.	1.752	1.738	1.755	Sobralite
432.	1.754	1.743	1.764	Caracolite
433.	1.755	1.736	1.766	Sursassite
434.	1.757	1.738	1.778	Piedmontite
435.	1.760	1.750	1.765	Nagatelite
436.	1.761	1.738	1.782	Manganepidote
437.	1.762	1.742	1.776	Gummite
438.	1.763	1.729	1.780	Epidote
439.	1.768	1.744	1.788	Diopside-acmite
440.	1.768	1.760	1.768	Corundum
441.	1.768	1.742	1.787	Aegirite
442.	1.769	1.702	1.796	Oxyhornblende

Table II–4 (continued)
List of Biaxial Negative Minerals

Code number	β	α	γ	Name
443.	1.769	1.760	1.769	Calcium larsenite
444.	1.770	1.745	1.782	Aegirite, vanidiferous
445.	1.771	1.751	1.782	Leucophoenicite
446.	1.771	1.728	1.800	Brochantite
447.	1.775	1.772	1.777	Britholite
448.	1.778	1.748	1.792	Hortonolite
449.	1.779	1.760	1.779	Allactite
450.	1.780	1.763	1.812	Diopside-acmite
451.	1.780	1.756	1.792	Alleghanyite
452.	1.782	1.758	1.804	Roepperite
453.	1.786	1.759	1.797	Tephroite
454.	1.792	1.768	1.803	Hortonolite
455.	1.793	1.756	1.809	Thortveitite
456.	1.799	1.762	1.814	Acmite (32 % Fe_2O_3)
457.	1.804	1.778	1.815	Tephroite
458.	1.805	1.78	1.82	Tephroite
459.	1.807	1.769	1.822	Acmite
460.	1.807	1.793	1.809	Sarkinite
461.	1.810	1.787	1.816	Arsenoklasite
462.	1.815	1.775	1.825	Pascoite
463.	1.816	1.776	1.836	Acmite
464.	1.822	1.786	1.833	Ferrohortonolite
465.	1.825	1.788	1.830	Hancockite
466.	1.831	1.800	1.846	Higginsite
467.	1.836	1.805	1.846	Manganfayalite
468.	1.838	1.805	1.847	Knebelite
469.	1.838	1.809	1.859	Linarite
470.	1.840	1.733	1.845	Chalcosiderite
471.	1.84±	1.80±	1.85±	Knebelite
472.	1.842	1.825	1.857	Dietzeite
473.	1.846	1.792	1.864	Iddingsite
474.	1.850	1.847	1.850	Romeite
475.	1.861	1.831	1.880	Atacamite
476.	1.864	1.824	1.875	Fayalite
477.	1.870	1.670	1.895	Tyuyamunite
478.	1.875	1.655	1.909	Malachite
479.	1.877	1.835	1.886	Fayalite
480.	1.895	1.78	1.92	Tyuyamunite
481.	1.925	1.750	1.95	Carnotite
482.	1.927	1.75	1.965	Tyuyamunite
483.	1.930	1.886	1.939	Fersmanite
484.	1.961	1.947	1.968	Alamosite

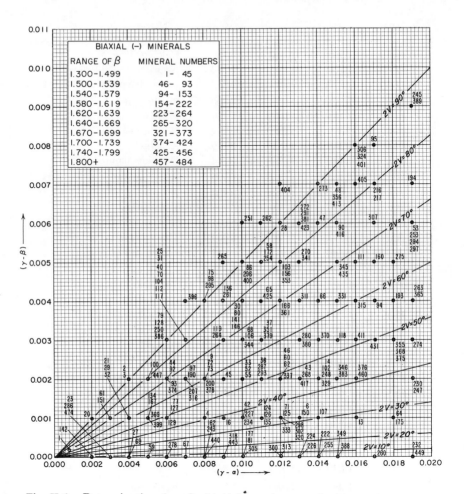

Fig. II-6. Determinative chart for biaxial negative crystals, $(\gamma - \alpha)$ from 0.000 to 0.019. Consult Table II-4 for identity of code numbers used in chart.

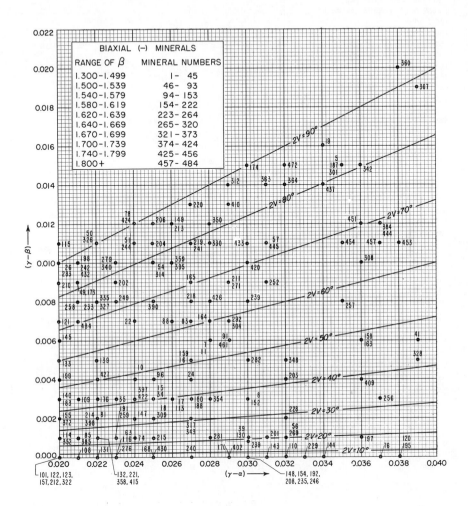

Fig. II-7. Determinative chart for biaxial negative crystals, $(\gamma - \alpha)$ from 0.020 to 0.039. Consult Table II-4 for identity of code numbers used in chart.

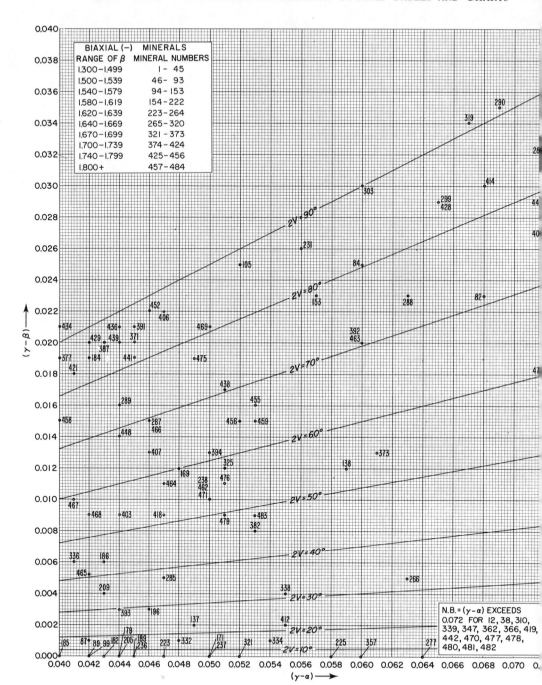

Fig. II-8. Determinative chart for biaxial negative crystals, $(\gamma - \alpha)$ from **0.040** to 0.072. Consult Table II-4 for identity of code numbers used in chart.

APPENDIX III ___ Recording Data

The form on the next page provides a convenient method for quickly recording measured optical data and serves as a checklist on the observations to be made. Data for a particular mineral are recorded in only one of the three blocks heavily outlined, the left-hand block being reserved for isotropic, the central block for uniaxial, and the right-hand block for biaxial minerals. By way of example, the sequence of measurements for a biaxial mineral has been entered into the right-hand block of the form. In addition to the data within the block, the macroscopic observations have also been recorded in the spaces provided. The column for the specific gravity of the oil can be used if the range in specific gravity of the oils is known. A downward arrow in this column indicates that the grain sank in the oil and therefore has a higher density; an upward arrow indicates the opposite.

MACROSCOPIC RECORDS

Reflected colors:
  a) of crystal:
  b) of streak or powder:
Specific gravity:
Hardness: Cleavage:

Sketches of oil mounts on slides:

☐ ☐ ☐ ☐ ☐
(1) (2) (3) (4) (5)

☐ ☐ ☐ ☐
(6) (7) (8) (9)

MICROSCOPIC RECORDS

No. in sketch	Data for oil in mount			Isotropic	Anisotropic				
	Index on label (25°)	Sp. Gr. of oil	Temp. of mount	n	Uniaxial: +, —		Biaxial: +, —		
					ω	ε	α	β	γ
(1)	1.550		23°		all much greater (>>) than oil				
(2)	1.650		23°				>>	>>	>>
(3)	1.700		23°				<	<	>
(4)	1.695		23°				<	>	>
(5)	1.695+1.700		22°				<	Match	>
(6)	1.690		21°				Match		
(7)	1.730	3.20	22°					<	
(8)	1.715	(grain)	23°						>
(9)	1.720	(sinks)	23°						Match

Index of matching oil, uncorrected:	1.690	1.698	1.720
Index, temperature-corrected:	1.692	1.699	1.721
Transmitted color:	Pale green	Brown	Green
Relative light absorption:	Least	Most	Moderate

Angles $2V$ and $2E$ as determined with Mertie nomogram from temperature-corrected, measured values of α, β, and γ: $2V = 58°$ $2E = 111°$

Angles $2V$ and $2E$ as measured from interference figures using Kamb, Mallard, or Tobi method (state which): $2V =$ $2E =$

Observed twinning: none

Inclusions (orientation and type):

Cleavage or fracture: Prismatic

Sign of elongation of crystals or fragments: +

Type of extinction   a) Parallel: b) Symmetrical: c) Oblique (max. angle): $Z \wedge c = 40°$

Dispersion: $r < v$ Symmetry of dispersional fringes along isogyres in interference figures:
$r > v$

a) Two planes (orthorhombic)   ☐
b) Inclined (monocl., 2-fold $= r$)   ☐
   Crossed (monocl., 2-fold $=$ A.B.)   ☐
   Parallel (monocl., 2-fold $=$ O.B.)   ☐
c) No planes (triclinic)   ☐

Mineral identity: #319, Salite

Fig. III-1.   Form for recording data. The data obtainable from crushed diopside (variety, salite) are written in to illustrate the use of the form.

INDEX

288